Agricultural
Choice
and
Change

PEGGY F. BARLETT

Agricultural Choice and Change

DECISION **Change** MAKING IN A COSTA RICAN COMMUNITY

RUTGERS UNIVERSITY PRESS
New Brunswick, New Jersey

Library of Congress Cataloging in Publication Data

Barlett, Peggy.
Agricultural choice and change.

Bibliography: p.
Includes index.
1. Agriculture—Economic aspects—Costa Rica—Case studies.   I. Title.
HD1802.B37   338.1'097286   81-13835
ISBN 0–8135–0936–X   AACR2

Contents

Figures

Tables

Acknowledgments

This research was made possible by a research fellowship from the National Institutes of Mental Health (Grant #1F01MH54608-01) and was aided by grants from Carleton College and Emory University. I would like to acknowledge my gratitude to those institutions and to Professors Conrad Arensberg, Frank Cancian, Bette Denich, Michael Harner, Marvin Harris, Allen Johnson, and Andrew Vayda. My debt to the Department of Anthropology at the University of Costa Rica, the Ministry of Agriculture of Costa Rica, the Consejo Nacional de Producción, and the Agricultural Extension Service of Puriscal is very great, and they have my sincere thanks. I would also like to acknowledge the support, encouragement, suggestions, and criticisms of my friends and colleagues Jane and Chris Baker, Nancy and Bob Hunter, Setha Low, Anne Farber, John Struthers, Michael Chibnik, Judith Tendler, Gordon Nelson, Peter Brown, Charles Melville, and anonymous reviewers. For work on the credit data in Chapter 8 I am grateful to Timothy Mitchell, and for invaluable help in my statistical analyses my thanks to Steven Moffitt and Michael Chibnik. Though the book is much better for the aid of all these people, its remaining shortcomings and errors are mine alone. My greatest debt is to the people of Paso, whose cooperation made this research possible and whose affection made this study such a pleasure.

Introduction

Paso, a small, formerly isolated farming community in the mountains of western Costa Rica, represents in microcosm the massive changes affecting many rural areas throughout the world. In this study, I hope to show the dynamics of rural change in Paso:

How the traditional slash-and-burn production of corn and beans has given way to the production of cattle for export to the United States and to intensive tobacco terraces that quadruple the farmers' labor costs.

How these two quite opposite changes in land use have acted both to worsen the already difficult situation of land shortage for many families and to ameliorate it for others.

How the formerly quite egalitarian frontier ethos of the community is challenged by the scarcity of land and the rising standard of living and how relations between wealthy and poor families—who are often kin—are changing.

How the community has responded to both new roads and new market forces from the outside and also to new government programs, especially agricultural credit programs. How the effects of these programs and agricultural changes both strengthen and weaken the egalitarian traditions in Paso.

In short, how Pasanos are responding to an ever shrinking world in which their population is growing rapidly, their subsistence decisions must take account of world fertilizer supplies, coffee prices, and national policies of beef and tobacco imports, and their hopes for new consumer goods, more education, and a different life are both nearer and farther from reality than they were twenty years ago.

In addition to exploring the changes in Paso, I wish to find out why they have come about. Working from an anthropological perspective, I see processes of change as new choices, behaviors, and attitudes on the part of farmers and their families. I have tried wherever possible to measure exactly what is going on and thereby to verify both my own and Pasanos' interpretations of current trends. My sense of Paso is not limited to what is happening "right now" but rather includes the forces that led it to be as it is now. By tracing these pressures on the community, I hope to show which ones will continue and how the future will be affected by them as well. Finally, I have tried to go beyond the simple determination of a list of factors that are important for understanding these changes in Paso and to specify *when* they are important, for which groups of Pasanos, and with what results.

Studies such as this one have direct relevance in solving practical problems in developing countries. As the gap between the rich nations and the poor nations continues to widen, and conventional approaches to increasing rural incomes have not been successful, the awareness is growing of a need to understand the internal structures of rural communities. National policies of population planning, agricultural development, and rural income generation need to be based on careful assessments of the current situation and the change processes already under way. In Paso, for instance, the real incomes of some families are declining not only in relative terms but in absolute terms, and government efforts to reverse this trend must begin from a clear analysis of what factors are responsible for it. Furthermore, as our global food interdependence increases, and experts continue to predict an impending world food crisis, our attention turns to the small farmers in developing countries whose productivity is not keeping up with their rising numbers. This potentially catastrophic situation can be ameliorated only by programs based on good analyses of what causes these trends in the first place.

This study of Paso will address four general areas of topical and theoretical concern. First, Paso can be seen as an example of agricultural evolution, of rapidly changing ecological adaptation to a changing international and external environment. Second, by unraveling the dynamics of agricultural production in the community, we can begin to understand the economic processes behind them. From this perspective of economic anthropology, the production decisions of each household can be examined to understand the micro-level choice process that makes up the larger macro-level changes. Third, Paso's experience provides some unique insights into the process of "development." Standards of living in many Pasano households are rising, in sharp contrast to many communities in developing countries; in other ways, however, the increasing economic marginalization and social polarization of some groups in Paso are more typical of the transformation of isolated rural villages. Finally, this study of a Costa Rican community fills a gap in our understanding of open peasant communities in Latin America.

ECOLOGICAL ADAPTATIONS

Ecological concepts can be used to understand agricultural and social change in Paso, although the scale of the unit being studied is different from the traditional closed ecosystem. As Netting points out, "The method of cultural ecology has been used with conspicuous success in studying peoples with a simple technology whose dependence on their environment is immediate and compelling. . . . It is perhaps less obvious that the ecological approach will yield equally useful results when applied to agricultural societies" (Netting 1968:19). Because the changes affecting Pasanos may come from such international issues as the global fertilizer supply and the value of the U.S. dollar, or from national pricing and credit policies, any notion of a bounded ecological system cannot be applied. Instead of attempting to define all the flows within a specific system, this analysis will focus instead on the adaptive strategies of farmers as they use their natural environment to try to meet their subsistence needs (Bennett 1976; DeWalt 1979b; Whitten and Whitten 1972).

Another issue is whether the research focuses on static or dynamic adaptations. Some traditional ecological studies have looked primarily at the accommodation of one group of people to relatively constant environmental circumstances (Conklin 1957; Netting 1968; Rappaport 1968; Vayda 1969; Lee 1979). Others, including this one, have emphasized the cultural adaptations to *changing* environmental conditions (Collier 1975; Durham 1979; Forman 1970; Hanks 1972; Netting 1969; Pelto 1973; Wilkinson 1973).

As in Geertz's pioneering work in Indonesia, *Agricultural Involution* (1963), human-land relations are shown in this study to be a critical parameter within which these agricultural adaptations must operate. Paso's population, like that of the rest of Costa Rica, is growing rapidly, and farming practices are changing in response. Theories of the role of population pressure in agricultural stability and change are useful in clarifying the dramatic changes in Paso (Basehart 1973; Boserup 1965; Carneiro 1961; Clark 1967; Clarke 1966; Dumond 1965; Harner 1970, 1975; Netting 1969; Spooner 1972).

A number of recent works, however, have stressed the importance of going beyond the traditional delineation of homogeneous ecological patterns to see the diversity of different groups of individual actors (Brush 1977; McCay 1978; Moerman 1968; Richerson 1977; Strickon and Greenfield 1972; Vayda and McCay 1975). By specifying the heterogeneous subsistence adaptations within the community, this analysis can move closer to specifying the causal variables and individuals' responses to them (Barth 1967; DeWalt 1979a). To study the role of population pressure in Paso, the existence of private property must be taken into account, a fact which changes the notion of population pressure from one facing an entire group equally to one facing certain sectors differentially. In this way, access to land and the internal differentiation in peasant communities are integrated into the ecological analysis (Beals 1974; Cancian 1972, 1979; DeWalt 1979a; Halperin and Dow 1977; Pelto and Pelto 1975; W.R. Smith 1977). I will try to link the stratification in the community and the effects of land scarcity to the agricultural choices of farmers, the ecological results of their choices, and the implications of current adaptations for the future viability of these choices in Paso's mountainous ecosystem.

The ecological perspective allows the Pasano data to address the issues of agricultural evolution and the causes of intensification. The increase in tobacco and grains production in Paso can be used as an example of agricultural intensification and suggests that labor efficiency and soil depletion may be closely connected to farmers' willingness to intensify. Exploring the usefulness of this Costa Rican situation for understanding general processes of agricultural evolution demonstrates the importance of linking household decisions and the factors that affect them to the macro-level theories.

In addition to the role of population pressure in the intensification of agriculture, some authors suggest that a desire for prestige and new consumer goods can also be involved (Pospisil 1963; Rawski 1972; Wilkinson 1973). In the Pasano case, these factors do operate for a specific sector of the community. Thus, an ecological perspective on agricultural change leads to a better understanding of farmers' decisions, the environmental impact of those decisions, and the role of population pressure and new consumption demands in intensification and agricultural evolution.

AGRICULTURAL DECISION MAKING

Population density, a central issue in many of the works noted previously, is like many ecological relationships whose character can be studied in economic terms by looking at decisions involving the allocation of scarce resources. The difference is often one more of scope than of substance (Brush 1977:17). The substantivist and formalist perspectives of economic anthropology are combined in this study through a focus on economic production, though distribution and consumption are also discussed as they relate to production decisions (Cook 1973; Orlove 1977a). My perspective in studying household adaptive strategies is formalist in its view of resources and microeconomic decisions, but these choices are linked to the institutions and patterns of life in Paso that are traditionally substantivist subject matter. The emphasis on diversity within the community, however, requires a micro-level analysis of individual decision making, rather than a more global description of instituted processes (Cancian 1972; DeWalt 1979a; Greenwood 1976; Halperin 1977; Moerman 1968).

The primary economic activity in Paso is farming, and various aspects of agricultural production in Paso are new: export markets, fertilizer and other inputs, new seeds and crop varieties, and new agronomic techniques. Many of these adoption decisions are linked to a certain crop, and thus the choice of land use is the primary issue, the other changes being aspects of that primary choice. My analysis of agricultural change therefore focuses on land use choices and looks at farmers' criteria in selecting a crop mix.

In addition to talking with farmers about their farming decisions, I also attempted to measure and compare the various crop options using my own etic categories (Harris 1968). My use of emic and etic measures reflects a recognition of my own and Pasanos' fallibilities as observers and of the necessity for some kinds of objective "checks" to verify our subjective interpretations (DeWalt 1979a; Johnson 1978). Usually these measures were congruent, giving two kinds of evidence of what was going on and why. In a few cases, though, what Pasanos described as the reasons for their actions were not substantiated by my etic measures. These cases support Netting's conclusion that such discrepancies between farmers' cognitive processes and their behavior are not problematic so long as the behavior elicited is appropriate to the well-being of the actors (Netting 1974:46). The differences, however, are significant for the analysis of the major variables influencing these agricultural decisions. For example, in relation to the decision to choose tobacco, a labor-intensive crop, several farmers said their large families were an important consideration. In fact, farmers with families of all sizes choose tobacco, and the lack of a large or even a medium-sized family cannot be shown to deter any farmer from selecting this crop. The emic evidence that farmers take into account their family size is not wrong; they undoubtedly do weigh this factor. Rather, their analysis of the key variables in the tobacco decision is not supported by the behavior of other households in this regard.

Although I have tried to verify the emic data I received with etic measures, I have not tried to use economists' or accountants' tools such as the production function (Finkler 1979; C. Gladwin 1979) or the internal rate of return (Acheson 1980; see also Greenwood 1976). Nor have I attempted to model farmers' decisions with linear programming or some other mathematical technique (H. Gladwin 1975; Johnson 1980). Such analyses can be very interesting, but they seek mainly to determine the extent to which computers can simulate human cognitive functioning, or to evaluate the extent to which farmers' decisions conform to certain standards of "good business." Neither goal is appropriate here, since I do not attempt to "get inside Pasanos' heads" to figure out *how* they make decisions (unlike C. Gladwin 1980; H. Gladwin and Murtaugh 1980; Ortiz 1980); nor am I interested in *evaluating* their choices. Instead, my goal is to determine the patterns of farmers' decisions and the variables that can be used to understand them (Chibnik 1978).

Household decisions are shown in this study to be influenced primarily by the resources, both land and labor, available to the family. Within the stratification system that affects farm decisions, certain choices are constrained by the land or labor that the household can bring to its production process. Mechanisms for renting land and hiring labor are also important adjuncts to family composition and land ownership. Farm decisions are complex for most groups of Pasanos; the evidence shows that a variety of factors, including profits, risk, labor and capital requirements, and yields, affect the decisions of each sector of the community. When possible, I have attempted to demonstrate where each of these aspects operates most forcefully to structure the economic decisions in the community.

RURAL DEVELOPMENT AND THE GREEN REVOLUTION

A number of the changes occurring in Paso relate to issues raised by other researchers about the effects of economic development in rural areas and the impact of new agricultural technology. The introduction of new crop varieties, fertilizer use, credit programs, and production for export are ways in which agricultural change in Paso is similar to the Green Revolution programs in Asia. Several adverse consequences of the Green Revolution have raised questions about the social and political results of this rapid technological change (Griffin 1974; Paddock 1970). The holistic approach of the anthropological community study is particularly well suited to pursuing the question of whether these same dislocations are a part of agricultural development in regions other than Asia.

The analysis of change in Paso can address several of the findings from Green Revolution research in various parts of the world. First, in many cases, studies showed that rural stratification was increased rather than diminished by the new agricultural methods. Larger landholders were able to adopt the new methods more easily, because of both their greater wealth and their access to scarce inputs and government programs, especially credit (Franke 1974:88; Ladejinsky 1973:137). Upper strata in many rural areas benefited disproportionately from the new techniques (Azam 1972:51; Beals

1974; Falcon 1970; Griffin 1974); in a few cases the overall standard of living of the majority of the population declined, often due to the effects of inflation (Amjad 1972:24; Frankel 1969, 1971). Some tenants saw an erosion of their position (Amjad 1972:37; Griffin 1974:75; Ladejinsky 1973:137; Mencher 1978), while large landowners often increased their holdings through consolidation of smaller plots (Jacoby 1972:69; Mencher 1978; Skorov 1973:17). Mechanization led to labor displacement in some regions, threatening to increase out-migration to the cities (Beals 1974). Polarization of social classes and rural unrest were noted in several Green Revolution areas (Griffin 1974:76; Harris 1972:30; Munthe-Kaas 1970a, 1970b; Wharton 1969:468). These social dislocations were paralleled by a series of ecological problems involving the adaptability of the new seeds to each micro-environment, their nutritional content, and the dangers of widespread use of a single seed variety. Each of these trends will be examined for Paso, and I will show that although the forces affecting this Costa Rican community are typical of many developing countries, the emerging patterns show some important differences with the results of Green Revolution researchers.

Three aspects of rural change in Paso are often seen as evidence of "development." The most noticeable of these, changes in standard of living, involve new kinds of housing, changes in diet and dress, and the advent of new all-weather roads, regular bus service, and electrification. Pasanos now also have increasing contact with many outsiders, such as: the agricultural extension service and, through this service, the university research establishment; banks and credit programs; church and educational institutions; and new government programs of many kinds. Linked to these contacts are the changes in agriculture that I have already noted. Many of these new aspects of daily life can be seen as solutions to new problems or as attempts to find new ways to meet needs that can no longer be met in the traditional fashion. As Wilkinson points out, it is difficult in some cases to label these new solutions as "progress" (Adams 1970; Wilkinson 1973), and some observers may want to evaluate their effects quite negatively. Many of these changes are bringing Paso into a more Western and commercialized way of life. To avoid the suggestion that my own ethnocentrism causes me to evaluate such changes as unambivalently favorable, I often use quotation marks when I refer to them as "development."

Development also implies increasing dependency (de Janvry 1975; Frank 1967, 1969; Griffin 1969; Wilber and Weaver 1979). The economic influence of the United States is a backdrop to many of Pasanos' agricultural adaptations. This study can be seen to address the challenge posed by C. A. Smith (1978) that more attention needs to be paid to the strategies used in the periphery for survival. Here, it is important, as noted, to distinguish the diversity of responses within the community: there is no one strategy for survival in Paso. Instead, the decision-making processes of individual households, structured by their access to land and labor resources, result in different strategies, with different impacts on the choices of others (Berry 1980). In some aspects, farmers in Paso are now more dependent in ways that are clearly disadvantageous to them; in other aspects, changes that bring dependency also bring upward mobility, a higher standard of living, and greater economic security (see Orlove 1977b). Thus, development theory as well as theory in ecological and economic anthropology is explored and tested below with data from Paso.

A LATIN AMERICAN COMMUNITY

Since Redfield's pioneering work, community studies in Latin America have contributed to a number of anthropological concepts used in peasant studies world-wide: the dyadic contract; image of limited good; the culture broker; patron-client relations; the rural-urban continuum; and social race. In this literature, however, lowland peasants have received much less attention than highland Amerinds; this description of Paso helps to round out that imbalance. Through its Spanish heritage and colonial history, Costa Rica shares much of the Latin American tradition. It is also of particular interest to Latin Americanists as part of the less thoroughly studied "Euro-American" area (Service 1955). The cultural patterns I describe represent the adaptation of Spanish culture to the New World, separate from Aztec or Incan empires and African traditions. As such, this account provides an important contrast to the life of Guatemalan, Peruvian, or Brazilian peasants and adds to the scarce literature on Costa Rica itself.

Whether Pasanos should be called "peasants" or "farmers" is not a meaningful distinction for the kinds of questions discussed here. Wolf's definition of peasants as agriculturalists in effective control of their land who aim at subsistence but not reinvestment (Wolf 1955) is useful as long as investment is not defined as buying more land. Pasanos, like most peasants, do invest cash in more land, but this practice is aimed not at removing them from agriculture but at enlarging the basis for subsistence activities and lessening the intensity of labor required. I find no substantive differences in the behavior of Pasanos and farmers in other countries; the differences are a matter of degree, not kind (Cancian 1972:189; Ortiz 1967:194). The cultural and natural environments in which farmers and peasants operate are different (Sahlins 1964:136), but their responses can be understood with similar perspectives. Thus, I have used *peasant*, *peasant farmer*, and *farmer* interchangeably with *agriculturalist*.

Paso provides insight into several of Wolf's generalizations about the open peasant community (Wolf 1955). The atomistic relations between households and weak mechanisms of community control are one such theme. In addition, Paso is characterized by egalitarian cooperation in some spheres and inequality and stratification in others. The tensions between these conflicting characteristics broaden our understanding of life in such open communities and illuminate the ways in which patterns of life respond to the current social, economic, and political changes.

Paso can be seen in the middle of a continuum of peasant/farmer commercialization, much more in contact with the larger national economy than are the Peruvian farmers discussed by Brush (1977) but still maintaining the integrity of its agricultural adaptations (unlike the Basque farmers described by Greenwood 1976). Pasanos are caught between a secure, low-income past and an uncertain future that may see them either wealthy or impoverished. As such, it is a prototype of the changing Latin American community.

METHODOLOGY AND DATA COLLECTION

My choice of Paso as the subject of this study was based on a desire to study the effects of changing agricultural technology among small farmers. Costa Rica is one of the only Latin American countries in which substantial change had occurred in these communities by 1972. Other areas of rapid rural transformation were primarily areas of large commercial farms rather than areas representing the kinds of villages in which most rural people live.

Within Costa Rica, I limited the choice of community to areas that combined subsistence grain production and cash crops. I excluded both areas of commercial cash cropping and isolated regions of predominantly subsistence production. Paso was chosen after visits to many suitable communities because it represented an average-sized rural community that had experienced, according to the agricultural extension service, significant agricultural changes for over five years (in India, social effects of agricultural programs were visible within two years, according to Frankel 1971). My own previous research experience in Costa Rica in 1968 helped in the choice of Paso and in interpreting the data.

This research was carried out from September 1972 to September 1973. Time references in the text ("ten years ago") refer to the ethnographic present, measured from the year of the research. During that year, I lived in the community and maintained an independent household, though I ate each day with a Pasano family. Research on national price and marketing structures was done in trips to the capital city; my residence remained in Paso. Pasanos are mainly found in their homes and fields, and daily visits to these areas resulted in the majority of opportunities for participant observation and discussion of the topics covered here. In addition to these unstructured situations, data were gathered in several open-ended interviews with each household head. I wished to gather information systematically from all appropriate Pasanos in order to avoid the biases of relying on "key informants" or especially congenial families. I also reconstructed genealogies and *compadrazgo* patterns for the community, and undertook the usual mapping and census. Direct measurements were made of agricultural patterns, including field sizes, yields, and labor input, and I attended and recorded most community meetings, masses, funerals, and so forth. Informal interactions in the general store, the local market town of Puriscal, and the capital city of San José added to my understanding about the community.

In gathering the detailed economic data on landownership, income, credit use, and other areas, my agreement with Pasanos was that these data would be used only in the aggregate, so that no one individual could be identified from the figures. To further safeguard the anonymity of informants, Paso and the names of individuals are all pseudonyms. I have tried, however, to translate their statements faithfully and to present here the patterns of their lives.

The methodology I used was influenced by both Arensberg and Johnson. Paso is studied in depth, in the tradition of anthropological community studies, but with a concern for its larger national and international framework.

> Though many authors have recognized in theory that their studies need to be related to a larger universe of social, cultural, and psychological phenomena, few studies have attempted to show how the larger society affects the community under investigation. [Arensberg 1954:119]

In addition, where possible, the quality of my own observations were tested and elaborated with quantitative measures of what Pasanos are doing and why (Johnson 1978).

OUTLINE OF THE BOOK

Chapters 1 to 3 describe the context of change and lay out the parameters of life in Paso. Chapter 1 presents the internal and external forces of change in the community, both past and present, and links those forces to the changing patterns of land use over the last twenty years. Chapter 2 looks at the life-cycle of the Pasano household and explores the institutions that structure the household's access to land and labor resources. The internal differentiation in the community is the subject of Chapter 3, which describes changes in stratification patterns, household consumption, and social, economic, and political relations between the strata.

Chapters 4 through 7 present the analysis of land use choices in Paso and their implications for several different theories of decision making. Chapter 4 describes the production techniques of the four crop options in the community and the fluctuations in their prices and marketing structures. Chapter 5 turns to the way Pasanos choose among those four options, and measures Pasanos' attitudes toward them as well as the differences among the crops, on a number of variables. Different strata assess their crop options differently, however, and Chapter 6 looks at the differences in land uses by stratum. Chapter 7 takes up the question of predicting land use choices and tests the theories of Boserup and Chayanov on data from the community. The results of this analysis are used to develop a flow chart of crop mixes in Paso, combining the simple, one-factor tests in the first part of the chapter to create a more complex analysis in the second part.

Chapters 8 and 9 complete the outline of changing agriculture in Paso, and explore the implications of these findings. Chapter 8 presents the rapid adoption of agricultural credit and traces the effects of these new programs on different sectors of the community. Chapter 9 applies the data from several chapters to theories of economic development, labor intensification, and agricultural evolution, exploring why Boserup's "law of least effort" seems to be violated by some farmers in Paso. The Conclusion presents a summary of the important findings of the preceding chapters and then discusses the implications of the Paso data for development programs and policies.

In sum, this study attempts both to analyze thoroughly the agricultural change in one community at one time and to place the community in the context of human history and the cross-cultural parallels of the forces of rural transformation. These two goals are compatible: the more accurate the understanding of specific decisions and behaviors, the more useful this example will be both for testing broader theoretical issues and for comprehending the global changes that link Pasanos and ourselves.

1

The Historical Perspective: Development, Population Pressure, and Land Use Changes in Paso

The context of rapid change in the rural areas of Costa Rica involves forces from both within and outside the community. Paso has experienced rapidly increasing contact with a wide range of external organizations and pressures. New forms of communication—radio and, recently, television—as well as a new all-weather road and regular bus service have allowed Pasanos to leave the community more easily and have allowed numerous outsiders to come in. New government programs, improvements in education and health services, and a wide range of changes in the community's agricultural economy are some of the results of this increasing contact. Many aspects of these changes could be labeled "development"; and thus Paso provides an important opportunity to explore the in-depth effects on all sectors of the community of a kind of development process occurring in many rural areas of the world.

At the same time, internal dynamics of change in Paso are working in their own ways to restructure the traditional adaptations of the farmers there. Paso has experienced a growing population, and in recent years this growth has created substantial pressure on its fixed land resources and on the fertility of its soils. Land scarcity from rapid population growth is further exacerbated by the concentration of much of the land in Paso in the hands of a small number of large landholders. The remaining land available to the majority of the community feels the pressure of increasing numbers more acutely than it otherwise would. This underlying stress of scarce land resources is increasing, and it permeates many aspects of Pasanos' daily life and agricultural adaptations. Even if the community remained as isolated today as it was fifty years ago, the ecological imperative of a growing population would demand a series of responses. Paso has not remained isolated, however, and the external forces of change have interacted with the internal demographic and ecological pressures, sometimes in concert, sometimes in contrast.

In addition to describing institutions and processes that characterize the community, it is important to convey some of the quality and texture of life there. To this end, I have abstracted three "themes" here to illustrate the principles and tensions of life in Paso. These analytical constructs are not a part of Pasanos' own interpretations but are useful to explore their activities, attitudes, and interactions. The reader familiar with other areas of Latin America will note that some aspects of life in this Costa Rican community are typical of other Latin American villages, but other aspects are very different.

The first theme can be labeled *atomism*. In a number of ways Pasanos are very individualistic and independent. They have few customs of corporate action, familistic loyalty, or group solidarity. Families remain isolated for much of their activities and respect an individualistic freedom of action. This aspect of life is reminiscent of other frontier and pioneer areas (Biesanz and Biesanz 1944; Margolies 1977); unlike the conflict-ridden atomism described by Lewis in Tepotzlan (1955), it can be summarized as "low cooperation, low conflict" (Romanucci-Ross 1973:47).

In contrast, some aspects of life in Paso form a thread of *cooperation* and *egalitarianism*. This theme involves a profound sense of the equality of all persons and thus expresses a willingness to work together under certain situations. Bennett (1968:301) notes a similar volunteerism and cooperation as typical of an egalitarian frontier society. This theme implies that all participants act in some fundamental way from an equal footing, and have some obligations to help each other out when needed. It contrasts with the intense familism described in many peasant communities (Banfield 1958; Foster 1967).

A third theme involves *inequality*. Differences in wealth and power, privilege, and authority exist in Paso, and these differences are carefully noted, at times avidly discussed, even flaunted. Stratification, both in access to land and capital and in consumption standards, permeates some of the activities and interactions in the community, while hierarchical relations with outsiders are also an aspect of life in Paso.

The inconsistencies and tensions between these three themes will be explored throughout the following chapters, from the perspective that these characteristics are changing as the lives of Pasanos are changing. The pressures, both internal and external, that are transforming Paso are also selecting differentially among the patterns reflected by these themes. Atomistic behaviors are in some ways challenged, in other ways reinforced. The egalitarian cooperation that characterizes some activities is steadily losing ground in the face of increasing stratification and inequality within the community.

THE COSTA RICAN SETTING

Archeologically, pre-Columbian Costa Rica is a fascinating contact area of tribes, chiefdoms, and states. The Chorotega and Nicarao language groups migrated south from the Mayan areas of Mesoamerica to settle along the Pacific coastal plain and the peninsula of Nicoya (Chapman 1958:11; Lange 1976). By the time of the Spanish conquest, there were eight or nine Chorotega towns, all of which had temples and markets (Nunley 1960:10). From northwestern South America came Chibchan-speaking tribes who settled in the Talamancan mountains and Caribbean coastal area. These slash-and-burn agriculturalists conformed to Steward's tropical forest type (Steward 1948 (5):669–772). Throughout Costa Rica, there were several Nahua settlements and Aztec ports of trade—separate enclaves of a more highly organized society within the tropical forest culture area (Chapman 1958:12). The tropical forest peoples probably migrated north from the Amazon around 3000 b.c., and the overall configuration of their way of life remained basically unchanged when the Mayan traders arrived during the ninth and tenth centuries a.d. (Chapman 1958:12, 165).

On his third voyage to the New World, Columbus landed in Costa Rica and gave it its name. By 1564, repeated colonization attempts were successful, and a permanent Spanish community was established at the present site of Cartago (Nunley 1960:14). Spanish settlements were isolated and almost self-sufficient, depending for their subsistence on corn, beans, sugarcane, wheat, and cattle. The type of agriculture practiced was an amalgam of Spanish and Indian techniques, and the "standards of living of the Spaniards and the Indians were probably lower than during the pre-colombian era" (Nunley 1960:14).

The absence of both Indian *encomiendas* and mines of precious metals attracted a different kind of Spanish colonist and required a different life-style from the *encomenderos* in Mesoamerica and the Andes (Stone 1975). The lowland tribes in the Meseta Central were decimated by disease and pushed into more isolated regions, and in general the Indians in Costa Rica were not successfully subordinated to the service of the Spaniards as Indians were in more densely settled areas (Seligson 1980). Costa Rica's democratic governments and absence of haciendalike social structure have long been cited as deriving from its early history as a poor colony of independent farmers (Biesanz and Biesanz 1944; Blutstein 1970; Hall 1978; Seligson 1980).

COSTA RICA AND THE CARIBBEAN AREA

Costa Rica's 19,695-square-mile area and 2.2 million people make it one of the smallest countries of Latin America, but it also shares many of the common characteristics of the region. Its Spanish-speaking population in the central highlands is predominantly of unmixed Spanish descent, like that of several other areas in Latin America, such as parts of Chile, Brazil, Colombia, and Argentina (James 1959). The population of Costa Rica, like many of its neighbors, is largely rural, and its major exports are coffee, bananas, and cattle. Its standard of living, however, is the highest in Central America. James notes that Costa Rica has the highest rural population density in Latin America—more than 1,500 persons per square mile in the Meseta Central, where roughly 60% of the population lives. Migration out of this area has been substantial but has not led to depopulation of the core, a characteristic shared with only three other areas of Latin America (James 1959:705).

The geography and climate of Costa Rica vary greatly—from hot tropical forests along the coasts to cold montane dairy-farming areas in the central highlands. Costa Rica's communities are also heterogeneous—banana plantations, coffee farms, cattle ranches, and peasant communities. "Costa Rica is a nation of small farmers" is a statement often accepted uncritically, for the past as well as the present. While census figures have been variously interpreted (Busey 1967; Denton 1971; Seligson 1975, 1980), data from my research tend to support the point of view that there is much stratification in Costa Rica, albeit hidden under an egalitarian facade. The proportion of peasant communities in the country as a whole is unknown, but there are three major areas of foodgrain production, combining cash and subsistence cropping in a traditional community setting. The Puriscal canton is one of these three areas, and Paso can be said to be typical of Puriscal communities, though it cannot be said to be typical of the country as a whole.

Paso—Community History and Setting

As Spanish settlements in the central highlands grew, they spread out to cover the floor of the valley. Shortly after Costa Rica obtained independence from Spain in 1838, settlers from the central highlands began to spill over the mountain ranges to the west (Nunley 1960:24 Seligson 1980). By 1860, migrants from a town near San José had penetrated the dense forests of the Puriscal area and come to settle in what is now Paso. There had been Indians living throughout this region, but with the influx of colonists, they moved west toward the Pacific. Descendants of the five families who founded Paso cannot recount any contact with Indian groups, though they cite valleys to the west that are reputed to be populated entirely by the Indians' descendants. Many Puriscaleños show a mestizo phenotype, but for census purposes they are all considered white. No Indian language or identity remains in Paso today.

Paso grew from the original five families to a population of 507 people and 75 households by 1972. The community spread over the top and sides of a mountain ridge, with creeks and property lines defining its boundaries with neighboring communities. Today, Paso comprises a total of 550 hectares (1,358 acres) and is a discrete unit of territory recognized both by the residents, who call themselves Pasanos, and by the Costa Rican government.

The "center" of Paso, at the highest point of the mountain, consists of the school, the community center, two general stores, seven houses, and a coffee-buying station. The rest of the households are dispersed, with some clustering by kinship ties. Most houses can be reached by a half-hour walk from the center of the community, but a few require an hour or more of climbing down the hillside. Paso's adjoining communities to the north and south have more concentrated village cluster settlement patterns, primarily because they have piped water systems. Since Paso is at the top of the mountain ridge, its households must depend on streams for their water supply in the dry season. None of the streams is big enough to supply more than a handful of houses, and therefore the houses are scattered all over the hills, to take advantage of dry-season water supplies wherever they exist.

The altitude at the center of Paso is 1,100 meters, and the climate is marked by two distinct seasons. The rainy season, from April until November, is cooler; it is sunny in the mornings and clouds move in from the Pacific around noon each day. Rains may last for an hour or all afternoon and into the night. The dry season may see a few rains, but for the most part it is clear and hot. Rainfall averages 108 inches per year. Daytime temperatures in Paso average 75°–80° F, but the variation can be considerable. The average humidity is 65%.

PASO AND THE OUTSIDE

Contact with institutions, programs, and life-styles outside the community has affected so many aspects of life in Paso that only a limited number of such areas can be explored here. One of the most significant changes in Paso in recent years was the building of an all-weather road by the Costa Rican government, ten years ago. Before the road was built, communication with the outside was much more limited. An ox-cart road connected Paso with its market town, Santiago de Puriscal; a trip to town in those days meant two or three hours on horseback, longer by ox-cart. Travelling merchants came into the community regularly to buy grains and cattle, and others came to sell to the general store and house to house. There were few radios in those days, and visits by priests and government officials were extremely rare. Pasanos were aware of the more important political issues of the day, but travel to town was infrequent, and some people had never visited Costa Rica's capital city, San José.

Once the highway went through, Paso was opened up to more contact with the outside. A bus service now runs through the three communities on the mountain ridge where Paso is located and connects them four times a day with the market town. Pasanos also have much greater contact with San José than they did in previous generations. Today, the trip to Puriscal takes half an hour, if the bus does not break down. From there, buses leave every hour to San José, and the ride takes nearly two hours. Round trip from Paso to San José costs ₡9—one and a half day's wages for a peon.

Nearly all Pasano households own a radio, which is often in use. News programs, sports, soap operas, and political opinion are all part of the ordinary person's contact with the world outside the community. Newspapers and books are rare, except for the almanac and mini-encyclopedia paperback sold by the school each year. Once electricity was brought into the community, several families purchased television sets, which have become the gathering points for large numbers of neighbor children each night.

To clarify the amount of contact Pasanos have with the outside, I asked all households the last two times the husband and the wife had each visited Puriscal and San José and the reasons for the visits. The intervals between the two dates provide a measure of the frequency of contact. This measure was carried out during the rainy season, in a time of neither above- nor below-average work load. Hence, it reflects a frequency of travel between the times of peak labor demand in the communty, when trips to the outside are curtailed, and the times of relative leisure and fiestas, when trips are more common. A visit to communities near Puriscal was considered a visit to that town, and the same is true for visits near San José. These two towns account for virtually all contact with the outside, however. Tables 1.1 and 1.2 show the percentages of households by frequency of contact.

Table 1.1 shows that nearly all male heads of households had been to Puriscal in the preceding week or two. Women had gone less often, though 39% had been there in the past week or two and another 36% in the past

TABLE 1.1. INTERVALS BETWEEN TRIPS TO PURISCAL

Interval between trips	Percentage of men (N = 44)	Percentage of women (N = 58)
1–2 weeks	91	39
3–4 weeks	4	5
1–2 months	4	31
3–11 months	—	10
1 year or over	2	5
No trips	—	10
	101*	100

*Numbers sum to more than 100% due to rounding.

TABLE 1.2.   INTERVALS BETWEEN TRIPS TO SAN JOSÉ

Interval between trips	Percentage of men ($N = 55$)	Percentage of women ($N = 63$)
1–3 months	42	35
4–11 months	13	17
1–2 years	33	22
3–4 years	5	10
5 years or over	7	13
No trips	—	3
	100	100

month or two. Contact with the capital city is also higher for men, but not as much as in the case of the market town. Forty-two percent of the men had gone to San José in the past three months, while 35% of the women had done so. One interesting finding is that 10% of the women said they had never been to Puriscal, and 16% of them said they had either never been to the capital city or hadn't gone in five years or more.

The reasons for the visits outside Paso revealed interesting patterns not only in the use of institutions outside the community but also in the roles allocated to husbands and wives. Men who customarily go to Puriscal every week or two (91%), go to buy food and supplies for the family. Women generally go to town for medical reasons, taking children to the hospital for illness, vaccinations, and the like. Women occasionally go to town for shopping, too, and in households where there are no men, women go regularly each week or so, taking up that role. Women's trips to San José are predominantly for medical emergencies and appointments (60% of all visits by women). This figure illustrates the range of low-cost health care services available in the capital city for rural families; San José hospitals receive referrals from all over Costa Rica to their outpatient clinics. Twelve persons in Paso go to hospitals in San José on a regular-appointment basis for psychiatric, medical, or prenatal care. In general, men go to San José for medical reasons (36%) about half as often as women. The rest of their visits are prompted equally by the need to do errands or to visit the bank. Women more rarely visit the city to shop or do errands. Both men and women go to the city to visit relatives less than one-third of the time. Visits relating to work opportunities are also infrequent.

Although Pasanos visit the city relatively frequently, many say they feel out of place there. Even relatively well-to-do men and women report feeling nervous in their dealings with many higher-status people outside the community. Some informants said they go to the city only because they have to, then hurry directly from the hospital or bank to the Puriscal bus. They cite the noise and traffic, the fear of getting lost, and the dangers of a city as reasons for their unease. In addition, rural people are often treated as subordinates by the shopkeepers and professionals they deal with, and these abrupt or discourteous interactions compound the tensions of the high cost of the trip. That travel outside Paso is so frequent in spite of these problems is a testimony to the value Pasanos place on the services available in the urban areas.

Most Pasanos feel more at ease about the costs and experience of going to Puriscal than San José. Even there, however, my observation of the interactions between some townspeople and poorer Pasanos occasionally illustrated the condescension and deceit so often described in peasant-town relations. Wealthier Pasanos are seen more as equals by townsfolk, and the quality of their interactions is more comfortable. Overall, dealings between townspeople and all rural folk tend to be amicable, if distant. Some storekeepers are friendly, the bank employees can be courteous, and the municipal employees are sometimes attentive.

Pasanos also leave their community for longer periods of work, and some have migrated permanently. Out-migrants are drawn from all strata of the community—landless families looking for work or land, educated sons and daughters of landholders of all sizes, and young men and women seeking temporary work or excitement in the capital city. (The proportions of out-migrants in Paso will be discussed later.) A number of Pasanos are return migrants, and they express a preference for life in the countryside. Stressing the economic benefits of being able to produce their own food, milk, and housing materials, they also point to the lower cost of living, slower pace of life, and greater safety of raising children in the rural areas as reasons for their return.

NATIONAL INSTITUTIONS OPERATING IN PASO

A range of outside institutions affect life in Paso, and have increased their programs in Paso in recent years.

Education

Costa Rica's reputation as a nation with a high literacy rate and a high regard for education in general is upheld in Paso. Almost all adults are literate, and of those over 15 years of age the average number of years of schooling completed is 4.0. Elderly Pasanos recount the days before the government supplied a teacher, when families cooperated to bring in a schoolmaster. The building of a permanent schoolhouse 20 years ago was a major event. Building materials were hauled in by ox-cart, and the community worked hard to see the three-classroom building become a reality. Staffed today by three teachers from the Ministry of Education (one of whom is the school director), the school offers six grades, up from the three grades available to the older generation. Classes meet for three hours a day (either morning or afternoon shifts) five days a week, and on alternate Saturday mornings.

Today, 91% of all school-age children are in school, illustrating the high regard people in the community have for education. Children are kept out of school by their parents for economic reasons ("not enough clothes") or because of physical problems (deafness, epilepsy), rather than from the need for the child's labor at home. Upon completing the six grades offered in Paso, any student who wishes may attend high school in Puriscal, but only two or three do so from each graduating class. There is no tuition for high school, but the costs of transportation, room and board, books, and uniforms make higher education very expensive. Further, since a high school diploma qualifies the student for white-collar jobs unavailable in Paso, higher education means out-migration as well. There is no case of a person from the community with a high school diploma returning to agriculture.

The school's activities are supported by two committees of parents, and both committees work closely with the school's director. The committees raise funds, discuss problems that arise in school, and stage several annual community events. These cooperative activities form a thread that runs through the otherwise individualistic ethos of schooling. Children bring home report cards and occasionally talk about lessons with their parents, but their progress is usually their own responsibility. One boy stopped going to school in fifth grade, and his mother shrugged. "He doesn't learn anything, so there isn't any point." I saw no evidence that any parents pressured their children for academic achievement, but a number took pride in that achievement, when students did well. I never heard comparisons made between siblings' academic ability, except to compare it with other skills, such as the case of one parent who said, "Alfonso is going on to high school; I'll miss him a lot and it will be hard not having him around. But he never was much good at farming anyway." Another mother said, "We've decided not to send Juan to high school; he loves working in agriculture and we really need his help on the farm. Besides, the costs are too high, and we have no security that he might get a scholarship."

The Catholic Church

Pasanos are all Catholics, and religion is part of the fabric of their lives, but attendance at mass varies greatly from individual to individual. Masses are held every Saturday in Puriscal, and a few people make an effort to go at least once a month. Others rarely attend mass in town but do go to the occasional masses in the neighboring community or in Paso. There is no church in Paso, partly because the community is so small and partly because the neighboring church is large and can serve the whole area. Pasanos are in the process, however, of building a chapel, part of a rivalry between the communities. At present, the priest comes from Puriscal irregularly to say mass in Paso.

Especially when the mass is in honor of a deceased Pasano, the majority of households are represented there.

Religious observances reflect the atomistic theme of life in Paso. Religious devotion or attendance at mass are considered a person's own business, and the size of the turnout for a mass is not a point of honor for the community or even a point of discussion for those who attend. Even parents take a relatively disinterested stance with their children, and I never heard any child being urged or coerced to accompany a parent to church. Pasanos do engage in fulfilling personal promises or pilgrimages, to repay the Virgin or a saint for the fulfillment of a request. One woman walked all the way to a shrine in Puriscal to thank the Virgin for the health of a family member.

The Catholic church supports the Society of Saint Vincent, a charity group under the supervision of the priest. The society works with a committee of volunteers and has given food and housing materials to a few of the poorest families in Paso. Some of the society's members have worked hard and selflessly to help others in the community; their behavior reflects the cooperative theme in Paso.

National and Local Politics

The three communities on Paso's highway elect one representative to the cantonal council, and for many years this position has been held by a Pasano. The delegate not only represents local interests in affairs of the town but is well known to the politicians and national political leaders there. Puriscal's representative to the national legislature comes to Paso about once a year to touch base with local people, as he does in all the Puriscal communities. His legislative appropriations were responsible for building the road, leveling a community soccer field, and extending the electricity line into the community. Community members discuss the ways their interests are tangibly served by this representative, and they do not hesitate to express to him their strong desire for a piped water system.

Paso has never had any local government, nor are there any autonomous local groups with political functions. While the community gains some clout with politicians from being relatively homogeneous in its political support of one party, it is not a cohesive interest group. When the electricity line went in, many people cooperated to give a party to inaugurate it, at which national and local politicians were thanked and new requests put forward. But the ethos of the event was "petition and gift" (Diaz 1966; Strickon and Greenfield 1972), and it was clear that power and authority rest with the central government, and local people have little control.

Agricultural Agencies

Contact with agricultural agencies varies in Paso, but at times it can be frequent. The agricultural extension agency from the Ministry of Agriculture has organized 4-S clubs, analogous to 4-H clubs in the United States. Young men and women as well as heads of households are members. The meetings usually take the form of talks by the agricultural extension agent on technical topics such as new corn varieties and vegetable growing, but they also have dealt with community development ideas from Pasanos. Agricultural extension personnel also come into Paso on occasion at the request of a farmer to give him personalized advice about a disease or pest infesting his crops. Officials from the Tobacco Defense Board and from coffee-purchasing companies also come in for meetings with growers and to give individuals advice during the growing season. Bank representatives come in to talk about new credit programs or to sign up participants, and a new tobacco-purchasing company held a meeting with Pasanos during the year of this research. Pasanos thus have a wide range of contacts with government and other officials, relating to their agricultural enterprises. All these visits are a far cry from the much-remarked visit 15 years ago of one extension agent who came in on horseback before the building of the road. Paso's accessibility by bus or car is essential to this close contact with outside programs and services.

Other Government Programs

Paso receives a range of visitors from other government programs. In the recent past, a mobile health team with a doctor, nurse, pharmacist, and community health worker made scheduled visits to provide low-cost medical care. The team also helped provide materials for the building of outhouses, a part of Costa Rica's national latrinization program. The Social Help Institute, an autonomous social welfare agency funded by the national government, gave housing materials and food to two poor families.

Surplus food from the United States, administered by Caritas through the Catholic Church, has arrived in Paso for many years. To get a portion of the food, families must either participate in the Women's Club (*Amas de Casa*), which learns crocheting and other domestic skills, or in the Men's Garden Club (*La Huerta Familiar*), which grows a collective vegetable garden. These groups have also received shipments of clothing from the United States. The *Huerta* is made up mostly of landless men, with one landed farmer as a leader. The *Amas de Casa* has women from all ranks of the community and is led by one of the schoolteachers. This committee structure was established by the Puriscal priest in charge of administering the Food for Peace program in the canton. His goal was to distribute food only to families who were working "to better themselves." All the households that received the food stressed its importance both to their diets and to their budgets.

For a period seven years before this research began, Paso had a policeman assigned to it. Although he lived elsewhere, he had an office in a small building with two wooden jail cells attached. After this time, another policeman was assigned to one or both of the neighboring communities. The functions of the *policía* depend greatly on the individual, but they include such tasks as the delivery of mail and of occasional messages from town. Particularly in the neighboring community, the policeman serves to keep the peace between drinkers at the stores there. The jail in Paso was used only once, to hold a drunk for the night.

There is also a soccer team in Paso. It travels to other communities in the region in competitive tournaments and receives these communities' players for home games. Both Peace Corps workers and 4-H Club students from the United States have visited Paso, though none has lived in the community.

General Characteristics

Meetings of formal organizations are usually held in the Community Center, a building partly financed by the Alliance for Progress and built with community help. It was constructed primarily to house the mobile health team, but since the team no longer comes to Paso, it continues to be used for masses, meetings, and celebrations. The Committee for Communal Welfare (*Comité Bienestar Comunal*) is in charge of regulating the use of the building and of raising funds for its improvement. Several groups, especially the committees for the school, organize *turnos*, or fairs, every year or so. A *turno* consists of several days of festivities involving the sale of food and liquor, live music or a jukebox, dancing, a soccer game, and horse races (which, incidentally, are nearly identical to horse races described for Navanogal, Spain, by Brandes 1973:753). The work of putting on a *turno* is carried out by the sponsoring committee, and the proceeds are used at its discretion; school improvements, furniture for the community center, and a religious statue are recent investments.

All three of the themes of life in Paso can be seen in these community activities. Participation and service on committees is voluntary, and the general attitude is that such work is a good thing to do, but only if one wants to be involved in that way. Families who do little for the general good are not criticized or censured. A certain amount of respect and prestige goes to the men and women who spend considerable time on community activities, and their influence sometimes carries over into political or economic realms. Obviously, Pasanos cooperate in this range of activities and work together to achieve certain goals, like the building of the community center. Unlike communities where family self-interest is the primary concern, in Paso service for the common good is considered necessary for the community to progress. Pasanos' cooperation, however, mostly links up with outside initiatives, and they often speak of it as "collaborating with this government program." At the same time, the egalitarian ethic in the community and the lack of a history of any local government lead to an awkwardness about community service. Heads of commitees seem uneasy with the dominant role of directing group activities, and most prefer leadership roles to be filled by high-status outsiders. There are tensions and confusions about the proper use of community property, and the rights and responsibilities of group members are discussed repeatedly in a fumbling manner. These aspects of their lives are new, and to a great extent, have been imposed from outside.

One example of the way the atomism of the community affects these activities is that each group struggles with getting messages to its members, and no easy communication system has yet emerged. Meeting times are decided on, then later changed by some of the leaders; some members are notified, others are not. The upshot is that people are rarely sure a scheduled meeting will be held, and members thus occasionally fail to show up because no one confirmed to them that the meeting was indeed taking place. Only the school director has been consistently successful in getting the school committees together; he types notices to each member and has them delivered by their children. Other leaders do not emulate his techniques, perhaps because they feel uncomfortable writing messages or lack the authority to ask children to deliver them. Further, not all members of other community organizations have children in school.

Both egalitarian and stratified themes run through the interactions of the members. When discussing activities and plans, differences in wealth tend to be ignored or considered irrelevant to the basically equal footing on which each person learns to crochet, goes to mass, or hears about a new bank program. On the other hand, plans to raise money or buy crocheting materials may founder from lack of support from poorer members, who cannot afford the expense. A number of poorer women either did not join the Women's Club or dropped out from lack of funds for dues. (Dues amounts were usually suggested by the schoolteacher and poorer women did not protest, but instead later dropped out.) Nominations to committees cut across wealth lines, but one wealthy woman smiled disdainfully at a landless woman who declined a nomination to the Society of Saint Vincent, saying she could not afford to give up another afternoon a week from the sewing she did to support her family. Another example of stratification in these activities is that each year for several years the *turno* to support school activities received the donation of a calf from one of the wealthiest families. The calf was then auctioned off as one of the activities of the *turno*. In general, however, raffles and dues are shared equally, by rich and poor alike.

The interactions with government leaders, professionals, and other outsiders who come into Paso for these activities are unambivalently hierarchical. Their social status clearly ranks above that of "agriculturalist," and their interactions reinforce this distance. Negative attitudes toward the country life and people are at times openly expressed, and otherwise take a more subtle form of condescension. Programs are usually presented in the form: "We want you to do this; are you going to cooperate or not?" Both speech style and dress set apart many of these persons, and Pasanos generally accept these marks of superior status as being appropriate. For example, one poor woman with neither running water nor electricity sympathized volubly with a schoolteacher who complained about her high monthly electric bill, due to her refrigerator, her electric stove, and her washing machine. Only once did I notice a farmer openly irritated by the condescension of outsiders, when he was belittled and addressed in the familiar verb form by a shop-keeper. (In general, the familiar forms of Spanish, either *tu* or *voz*, are not used in Paso. Parents, children, spouses, and neighbors all use the formal *usted*. Only once did I ever hear the familiar plural form *vosotros* used, when a schoolteacher was speaking eloquently to the women's club.) When I questioned another Pasano about why she had so warmly supported a proposed series of lectures that I knew she did not think would be very useful, she replied, "We may not get a lot out of this program, but if we don't cooperate, they won't come back again, and maybe the next program will be something we will really want."

POPULATION AND ECOLOGICAL TRENDS

Turning now to the internal processes of change, we can trace change in Paso's population and ecological adaptations. Population growth has been characteristic both of Paso throughout its history and of the rest of Costa Rica as well. Nunley has documented that the population of Costa Rica has nearly doubled every 33 years since the nation achieved independence in 1838 (Nunley 1960:49–65).

According to Costa Rican national census figures, Paso's population grew 30% in the ten years from 1963 to 1973. Going back another decade, however, figures for Paso alone cannot be isolated from the 1953 census of the region. Assuming the growth rate has been constant, extrapolation back to 1953 yields a 69% increase in Paso over 20 years—a doubling period of roughly 30 years. To check this extrapolation, the figures for dwelling *houses* for Paso are found in the 1953, 1963, and 1973 censuses. The 1953 census lists 42 houses in Paso in that year, compared with 57 in 1963 and 77 in 1973. The rate of increase in number of houses has thus been constant in the last 20 years, a doubling period of 24 years.

The evidence from the houses, then, supports the estimate that Paso's population has been doubling at least every 30 years. This pace of growth can be confirmed by the fact that the original five families of Pasanos settled there around 1860. With a doubling of population every 30 years, there should be 40 households in 1950, and the 1953 census lists 42 houses. This calculation shows that Paso participates in Costa Rica's nation-wide process of rapid population growth. While informants claim that infant mortality was much higher in the past, it is nevertheless clear that a higher growth rate is not a recent change in Paso.

To assess whether this population growth stems from natural increase or from migration, I recorded genealogies for all Pasanos, dating back to the original settlers. These genealogies were analyzed and five generations of cohorts were separated (these lines are inevitably arbitrary, as siblings sometimes span more than a generation in age difference). The first generation consists of the children of the original five couples who came to Paso. Table 1.3 shows the percentage of out-migration for each generation.

TABLE 1.3.   PERCENTAGE OF CHILDREN BORN IN PASO
             WHO MIGRATED ELSEWHERE

Generation	Number in generation	Number of out-migrants	Percentage of out-migrants
1	24	8	33
2	124	50	40
3	270	170	63
4	256	88	34
5	206	10	5

The low 5% out-migration rate for the children of the fifth group means only that this age-group is young—most of these children are not yet old enough to leave home. Except for the third generation, which will be discussed further, the out-migration rate is constant around one-third, confirming that although out-migration from Paso is significant, it has been relatively constant over time.

Migration *into* the community is also a continuing process in Paso. Immigration comes primarily from intermarriage with families in nearby villages, but there are also some cases of married couples or single men who have moved into Paso, buying land or working as wage laborers. These newcomers or their children almost always intermarry with the older residents. It is difficult to analyze genealogically the in-migration rate throughout the history of Paso, because although in most cases informants can be sure someone *was* born in Paso, they cannot be sure if some ancient relative came into the community as a child or his parents did. Therefore, a study of in-migration for the couples now living in the community must suffice. Since children cannot migrate into the community without their parents, and since there are no unmarried immigrants or unmarried couples in Paso at present, statistics of the married couples give an accurate picture of the proportion of newcomers. Using the same generations of cohorts as in Table 1.3, the data show that 30% of the spouses in generation three and 33% of the spouses in generation four are not native to Paso. These figures are so close to the

33%–40% rate at which Pasanos leave the community that the population growth in the community can be attributed primarily to natural increase. In fact, since out-migration exceeded in-migration by 33% in the third generation, the population growth rate is actually somewhat higher than the net increase in community size.

Since there is no written history of Paso, and most informants' memories are hazy, it is hard to reconstruct what factors may have led to the high out-migration rate of 63% in the third generation. It seems, however, that this generation came of age during a period when one man amassed a very large farm, at one point controlling more than half the land of the community. This land concentration may have created a shortage of land for new householders, encouraging them to look elsewhere. When Pasanos were asked why relatives left during this period, they most often replied, "He/she left in search of land." During this time also, the Costa Rican government passed a homestead act encouraging settlers to claim lands in the mountains west of Puriscal, and this opportunity may have drawn Pasanos. In any case, these opportunities declined in later years as land to the west became scarcer and as the heirs of the one large farmer sold some of his holdings and eased the concentration of land inside the community.

Only a few comments can be made at this time about the causes of this high rate of population growth (see Thein 1975). The community average completed fertility is 6.1 live children per couple, suggesting that fertility is very high and relatively unrestricted. There seem to be few customary checks on fertility of the sort discussed by Davis and Blake (1956): there are no postpartum or ritual taboos; frequent marital intercourse is considered a sign of virility; abortifacients are not used; early marriage for women has always been common; prolonged separation of husband and wife is rare; widow remarriage is permitted; and venereal and other diseases lowering fecundity are not common. While modern birth control methods have become available in the last five years, their effect on fertility has only begun. The high fertility rate seems to be an appropriate response to a frontier situation in which land is perceived to be abundant and labor is the key to building family wealth. As recently as 1944, Costa Rica was described as having "an abundance of land combined with a small population" (Biesanz and Biesanz 1944:152). Although the exact causes of the high fertility rate cannot be specified, it is clear that such a pattern is characteristic of all of Costa Rica and is not a recent development.

This process of population growth in Paso has led to changes in agricultural practices, resulting in shorter fallow periods and soil depletion, a sequence described by Boserup (1965) and others (Brookfield and Hart 1971; Gleave and White 1969; Haswell 1973; Clark and Haswell 1964; Knight 1974; Netting 1977; Prothero 1972; Redfield 1950; Ruthenberg 1968). The early colonists in Paso used slash-and-burn techniques to plant corn, beans, and rice, both for subsistence and for sale. Cutting a new plot each year, farmers then abandoned old fields to return to forest. One elderly Pasano said his father always made sure his plots lay fallow for five years before cultivating them again. Some households also kept some pasture for milk cows and planted a stand of coffee for additional income beyond grains. (Throughout the book, I will refer to corn and beans, and sometimes rice, as "grains." Though beans are not technically a grain, this shorthand follows Pasanos' own reference to *granos* as well as the usage of development programs aimed at "basic grains," both of which include beans.)

As population has increased and farms have decreased in size over the last century, fallow periods have been shortened and eventually abandoned. Boserup points out that "when fallow is shortened or even eliminated in a given territory, some other method of preserving or regaining fertility must of course be introduced" (1965:25). Without such methods, soil fertility drops, and often erosion washes away the soil altogether. "Barren hills deprived of their earlier vegetation and topsoil abound in most regions of ancient civilization" (Boserup 1965:21).

Paso is just beginning to face the problems of declining soil fertility. Practices such as composting or manuring are not in the agricultural repertoire, and farmers have turned instead to purchases of chemical fertilizers. Fertilizer has halted the decline in crop yields in many fields, but farmers testify that in some plots, corn will not even produce an ear without fertilizer.

> One farmer said, "The land is tired now. We used to get good crops, but now we get little, even with fertilizer."

> His friend added, "That's right. And we are trapped because the price of fertilizer keeps going up and up."

Is this loss of soil fertility due entirely to population pressure and the decline of the fallow period? Or were the Spanish descendants in Paso "inept" in their use of shifting cultivation? "Wasteful or inept methods may be destructive to the long-run equilibrium of swidden agriculture" (Geertz 1963:26). There is no way to determine the actual agricultural methods in use a hundred years ago, or to discover whether the productive system was in equilibrium with the environment or was slowly degrading it. There is, however, indirect evidence from Talamancan Indians who live on the southern border of Costa Rica. They practice slash-and-burn agriculture and get corn and rice harvests much higher than what Pasanos get per hectare. The Talamancans state that a fallow of five years is sufficient to maintain fertility, but their climate is slightly warmer and wetter than Paso's, and their soil is undoubtedly different also. Their use of a successful five-year fallow may be counted as evidence that five years may possibly have been adequate for Paso as well. The differences between the two ecosystems are sufficient, however, to ask for further proof. In any case, even if the original agricultural practices were ecologically sound, population growth over time reduced the fallow period to the point of soil depletion, and it is at this point in the history of Paso that my research was carried out.

LAND SCARCITY AND CHANGING LAND USES

The internal population growth just described intertwines with external market forces to change the land use choices of Pasanos. These crop choices affect, in turn, the availability of land for other farmers and the relationships between landed and landless Pasanos. They also have dramatic impacts on the ecosystem and the long-term viability of agricultural production.

The marketing mechanisms for the different crops grown in Paso will be discussed in greater detail in Chapter 4, but several general points can be made here. First, Paso was primarily a grain-producing community prior to the building of the highway. In fact, the entire Puriscal region exported corn, beans, and rice to the cities of the central valley. Cattle production was also common, from the days of the earliest settlers, but exports of beef were not a primary commodity until more recently. Coffee was also grown in Paso, but transportation difficulties inhibited its development as a major cash crop.

Since the building of the all-weather road, new markets and technologies have become viable. Tobacco has become an important new crop. Coffee plantings are on the rise now that the harvest can be easily transported, and a buying station has been built in the community. A sharp rise in beef prices and the expansion of the United States market for imported beef has led to an increase in land committed to pasture. Corn and bean prices have not increased as fast as prices of these other land uses, and with declining soil fertility, rice has been virtually abandoned as an important crop in Paso.

In order to measure the changing patterns of land use in Paso, each household was asked to draw a map of the pieces of land it owned, listing with whom the land has boundaries. Maps for the whole community were later matched to make sure no land was missed. Using the map as a reference, each farmer traced the history of land use for each plot. Twenty years' time depth was easy for almost all householders to remember, since the past use of the land is a factor which is discussed when land is bought. Furthermore, the happy coincidence of the construction of the school 20 years ago and the road 10 years ago helped informants remember what they had done with each plot at those times. In the cases of owners who were too young to remember what land had been used for 20 years ago, older Pasanos were able to supply the information.

Table 1.4 distinguishes seven different land uses. The agricultural methods involved will be discussed in greater detail in Chapter 4, but at this point brief definitions are necessary. Tobacco is always planted in rotation with corn and beans, so that the decision to use land for tobacco is actually a decision for all three crops. (I use the phrase "in rotation with" to express the multiple cropping pattern of annual tobacco production, in which tobacco occupies the field for several months, followed by several months' rest, then by corn and beans planted together). "Pasture" includes both improved and unimproved grasslands. The grains grown in Paso in 1972 are mainly corn and beans, but one or two farmers planted rice as well. This land generally supports two harvests in one year. "Fallow" land is usually in dense brush, and "forest" land has not been cultivated for many years. Coffee is a permanent crop, and except for fruit trees for shade, the land can be put to no other use. Miscellaneous crops are sugarcane, yuca, cabbage, broom corn, and fruit trees.

TABLE 1.4.   LAND USE OVER 20 YEARS, IN MANZANAS
             (.69 HECTARE; 1.7 ACRE)

Year	Tobacco with grains	Pasture	Grains	Fallow	Forest	Coffee	Miscellaneous	Total
1952	1	223	258	247	138	31	19	916
1962	18	301	198	226	101	49	24	916
1967	37	392	178	134	96	58	25	916
1972	45	535	96	94	63	71	14	916

Table 1.4 shows major shifts in land use over the past 20 years. Although the number of manzanas in tobacco is small, this new crop is one of the most significant changes in the Puriscal area. Pasture has more than doubled in the past 20 years, taking land away from grains, fallow, and forest. Coffee has also doubled in this period, while miscellaneous crops have stayed the same. It should be noted that the proportion of grain land to fallow 20 years ago is one to one, but this is an overstatement of the amount of land actually in grains in 1952. This distortion arises because 20 years ago several large tracts of land were in fallow, with only a fraction used for grains in any one year. Today these tracts are broken up in many pieces, but each landowner says, "This land was in grains 20 years ago," which tends to inflate the proportion of grains to fallow. This problem does not apply to the other categories.

Table 1.4 provides the diachronic data from which the following analysis of agricultural decision making begins. Two changes in land use stand out: the shifts to pasture and to tobacco. Tobacco is a very labor- and land-intensive crop, suitable for small farms with scarce land. Pasture, on the other hand, is a very extensive land use. Large landholders have taken lands out of fallow and forest and out of the hands of renters in order to increase their holdings in pasture. This decline in the availability of rented land sharpens the population pressure already evident on the lands available to the majority of the community.

The increasing scarcity of land has crystallized the tensions between the individualistic, egalitarian, and stratified themes of Pasano life. The predominant community individualism is seen most clearly in each farmer's autonomy in running his agricultural enterprise. For the most part, property rights are inviolate, and a farmer can do what he pleases with his land and other resources. In one case, a family had spent considerable effort and money to dig a well beside their house. Although they fear it will run dry once their neighbor cuts down his adjoining forest, they cannot protest his decision to use that land for crops. This individualism is bound up with the agricultural technology of the community, which requires virtually no interfamily cooperation or group effort at any point in the agricultural cycle.

At the same time, the growing shortage of land has led many landless farmers to criticize large landholders who hold land in fallow or pasture and refuse to rent it. From the day I came into the community until the day I left, I heard an endless refrain of the unfairness of the "rich" in refusing to rent their land. An egalitarian theme of the rights of all to a livelihood and a share of resources runs through this criticism. Fallow is no longer seen by some as a legitimate land use, when some families are undernourished from scarcity of land to rent. Thus, larger considerations of equity in access to resources challenge the traditional rights of the farmer to do with his land as he sees fit.

Large landholders occasionally take the position that the "poor" are lazy and don't know how to farm adequately. They are held to be deceitful and profligate, with the problem of poverty in their own characters and lifestyles, not in the shortage of land. This point of view denies the inherent equality of all Pasanos and justifies the unequal distribution of land. All three threads—individualistic, egalitarian, and stratified—can be heard from the same people at different times. In different contexts, a Pasano can affirm that each person has a right to farm as he or she wishes, a person has a right to sufficient resources to earn a living, and some farmers are poor from lack of skill or energy, not lack of resources. The growing shortage of land throws into relief the tensions between these traditional views, and the issues raised go far beyond the confines of this one community.

Other effects of these land use changes are ecological. Houses used to be isolated from each other, surrounded by tall forest. One old man complained:

One used to feel all alone on this hill, and now look—houses every-
where. It's a city!

As the forests are cut down, for pasture or for crops, houses now experience a
different climate. One family lost their forest protection when their neighbor
cut his woods across the road; now they complain of cold, damp winds that
beat on their house as clouds roll up the mountain from the Pacific. Other
households note that when the forests are cut, their water supply during the
dry season dwindles. Some creeks no longer last until the rainy season starts
again, and water shortages cause considerable hardships for women who
must haul water up steep slopes. Deforestation has also removed the animals
and plants that provide variety to local diets; especially for landless families
who must buy most of their food, the loss of such meat and fruits has a
significant impact on nutritional welfare.

     These ecological effects are not confined to Paso. Erosion from defor-
estation and the increase in pasture is a national concern. The hydroelectric
dam that serves the capital city has suffered from silting and inadequate
water supplies in the dry season. Although their observation is un-
documented scientifically, farmers in several areas of Costa Rica report that
the weather is changing. Cloud formation and rainfall patterns over the
denuded hills are different, and some areas report more frequent droughts.
Many of these changes would have come about eventually from the pressures
of increasing numbers on the land, but the new market for export beef has
greatly accelerated the process of deforestation, soil erosion, and ecological
change. Paso is thus a microcosm within which to study these internal and
external processes of change. We can turn now to the life cycle of Pasano
households and trace the effects of these processes on their traditional
patterns of access to resources.

2
Land, Labor, and the
Life Cycle of the Household

From many perspectives, the household is the primary unit in Paso. Economically, it is a cooperating group that makes joint production decisions. As such, the household is an economic unit of land, labor, and other resources whose goals are primarily to meet its own consumption needs. Socially, the household is the recognized unit of childbearing and care and of social interactions and obligations. It is also a unit in the eyes of the cantonal and national government, and it is seen as a moral unit by the church. To understand many of the processes of change in Paso, the life cycle of the household must first be explored. Since economic issues are the focus of this analysis, we will pay special attention to the household's access to land and labor resources.

COMPOSITION AND LIFE CYCLE OF HOUSEHOLDS IN PASO

There are 75 households in Paso, and 52 of them consist of an economically active married couple and their unmarried children. Another 2 households are elderly couples who are economically inactive and dependent on their children living nearby. Six households are extended by adding an aged parent or a sibling of the husband or wife. All 6 of these households function, however, as if they were nuclear families, bringing the total of essentially nuclear family households in Paso to 60, or 80% of the total. Of the remainder, 8 are widows with children (1 economically inactive), 4 are single persons (two cases of two brothers living together, one single woman, and one single man and his housekeeper), and 3 are extended families in which married children live with their still-active parents. Thus, the predominant pattern of nuclear family households leaves only a small proportion of never married and widowed persons, and only 4% of Pasano households have to deal with the more complex decision making of an extended family.

The establishment of a new household is the culmination of the process of courtship and marriage. Courtship in Paso follows the general Latin American pattern in which a man who is interested in getting to know a woman asks her to be his *novia*, or girlfriend/fiancée. If she says yes, they then begin to spend time together. Traditionally, the *novio* is supposed to visit his girlfriend at her home, but nowadays they can also go to masses or church meetings or even to 4–S club meetings at the community center. If they learn enough about each other to decide they are not interested in marriage, either party can break the relationship off. Once the couple has decided they wish to marry, the *novio* must ask the woman's parents for permission, and a date is set. The most common age of marriage for all Pasano couples is between 18 and 19 for women and 24 for men.

All couples in Paso were married in the Catholic church in Puriscal before living together, and in general, Pasano courtship follows the moral teaching of the church. Physical contact between *novios* is generally prohibited, but it is acknowledged that a few women have "had to get married" because of pregnancy. There have been a few illegitimate children born to unmarried women, but they total less than 2% of the current population. The traditional double standard of sexual behavior is accepted by most men and women, and it is assumed that young men will visit prostitutes or otherwise pursue sexual experience before marriage. A women who has sex with her *novio*, however, is felt to be risking the marriage, and one young man said "such a woman would be choosing the other profession of women," namely, prostitution. One woman in her early thirties said the first time she had ever touched her *novio* was when the priest joined their hands in the marriage ceremony and, she said, "I nearly fainted." Extramarital sex is as equally forbidden as premarital sex, but there seem to be no automatic consequences, and divorce is unknown. In general, Pasanos rarely discuss such peccadilloes outside the immediate family and consider it somewhat sinful to do so—an example of the atomistic and independent aspects of Pasano life.

The groom's family must bear all the expense of a wedding and of setting up a new household. By tradition and by practice, the groom must build the house, furnish it (including such items as dishes and bedding), buy his wife's wedding dress and shoes, pay the marriage costs, and give a party for relatives and friends. The girl's family may give a party if they wish, but they are under no obligation to do so.

Postmarital residence depends primarily on access to land and house sites. If a couple owns land, they usually build their first house on it, although if the site has poor water resources or access to roads, or no level land, some other site may be sought. If the couple is landless at marriage, they will generally ask permission from the husband's parents to build a house on their property, since the husband is more likely to continue to work on his father's land. There are many cases, however, of postmarital residence on land belonging to the wife's relatives, and it is clear that couples choose the location of their house according to the availability of land and their own convenience. Once the household has been established, the house site is rarely moved. Such a change of residence can be brought about, however, by the need to construct a new house or the purchase of new land with a more convenient house site located on it.

Children in marriage are considered to be a gift from God, and only recently have a few couples become interested in birth control (see Barlett 1980a). In general, families are large in Paso; the average household size is 6.6 persons. Infant mortality, according to my interviews with all mothers in Paso, is 160 per 1,000, but younger mothers have experienced a lower rate, due perhaps to new medical facilities in town and the new highway that allows them to get there quickly, or perhaps to the increased availability of modern medicines.

Remarriage after the death of a spouse is common in Paso. Widowers tend to remarry promptly, especially if there are young children in the family. Widows, on the other hand, are more likely to find new husbands only if they are young; older women generally remain with their children. A few elderly parents have abandoned their separate homes and moved in with their married children, but most Pasanos die in their own homes.

There are no cases of orphaned children, but all parents see godparent (*compadrazgo*) ties as providing for that eventuality. *Compadrazgo* ties are relatively unimportant in Paso to both parents and their children. *Compadres* are usually close kin—siblings, cousins, and even grandparents—regardless of whether they are for baptism or for marriage. Most parents ask the same couple to sponsor all their children, thereby greatly limiting the number of such ties possible. Though everyone I asked could always remember who their godparents were, there were no customary exchanges of gifts or favors between them, and I saw no evidence that economic cooperation was more likely between *compadres*. Godparents seemed to be designated primarily to fulfill Catholic doctrine and to assure that children, if orphaned, would be cared for. The contrast of this pattern with many other areas of Latin America further underscores the atomistic aspect of life in Paso.

When a Pasano dies, neighbors and relatives assemble in the home of the family and say a rosary around the casket. Visitors are usually served something to eat and drink, and then the crowd accompanies the casket to the cemetery, where it is buried without ceremony. Although the closest relatives may cry, public expressions of grief are not expected of them, and I did not see visitors cry at the two funerals I attended. The family arranges for a memorial mass to be said in Paso and arranges additional masses at fixed intervals, if they can afford to do so.

Armed with this brief overview of the life cycle of the household, we can turn to the acquisition of resources and their dispersion over the household's lifetime. Land, always a crucial resource for agriculture, becomes of paramount importance in situations of scarcity. Labor, the household's second most important resource, also passes through a process of addition and dispersal. Capital will not be discussed here; Chapter 8 is devoted to its use in agriculture and changes in that use in recent years.

ACCESS TO LAND

Rental, purchase, or inheritance are the three ways a new household can obtain land resources. Some peasant farmers do well in their agricultural enterprises from their early adulthood on and are able to buy pieces of land early in their marriage. Other households do not expand rapidly for many years, often until the oldest sons are out of school and working full time, around age 13. Still other households manage to buy a small plot of land and pay it off, but never do well enough to buy more, in spite of the labor of their children. In some cases, land is only obtained by inheritance; in others, the inheritance comes long after the farm has stopped expanding. Many landless families see little hope that their crops will ever produce enough of a surplus to allow them to buy land, so they depend on rented land for agricultural production.

Land Rental

Couples who do not own enough land seek land to rent. Regardless of whether the prospective renter is an unmarried man, a newly married landless man, or a landholder, the process of land rental is the same. The agreement to rent land is almost always oral, though there are a few cases in which the two men wrote the agreement on paper. The landowner specifies how much land, where, and for what crops he is willing to rent, and how much, in cash or in kind, he will charge. All costs for seeds, labor, fertilizer, and other expenses are borne by the renter. Rents vary greatly. Rents for corn and beans are usually expressed as so many units of grain harvest per unit of seed planted. Land rented for tobacco is almost always rented for a fixed cash price, but there is also a case of land provided free for the production of tobacco, with rent levied on the corn that follows in rotation.

As would be expected, rents are rising with the cost of land and with its scarcity; at the same time, the quality of the soils being rented has declined. This combination interacts to lower the profits obtained by many renters.

Compared to many parts of the world, however, rents in Paso are quite low. To combine rents in cash and in kind, the total cost to the renter can be expressed as a percentage of his profit. The profit used here is based on a Chayanovian calculation of the cash value of the harvest minus cash costs (see Chapter 5 for the details of this calculation). The average proportion of profits paid in rent in Paso in the year of this research was 22%. The lowest proportion of rent was 5% of profit; the highest was 71%—a case of harvest failure. The highest percentages of rent when the harvest was average or better were 47% and 50%. Since these are percentages of *profit* and not of the total *harvest*, it can be seen that rental patterns reflect a time of lower population density in Paso.

In bad years, landowners tend to be charitable in Paso. Several recent cases were recounted to me where no rent was charged or rents were reduced because of crop failures. In only one case was the landowner unwilling to forgive his fee, and in this case, payment was postponed for a year, until the next harvest.

Rental agreements are made for one year only, leaving landless families with the task of finding new plots each year. Some of this transience comes from changing land uses on the part of owners, especially from taking land out of agriculture and sowing pasture. One landless man listed five plots from five different landowners that he had rented in recent years, but all were now turned into pasture. Large landowners also say they prefer to avoid renting continuously to the same person because, under Costa Rican law, a tenant has claim to the land after working it for ten years. Landless families point out that no case has ever occurred in Puriscal of a tenant making such a claim. The pattern of shifting rentals is clearly more to the benefit of the landowner; renters would prefer the greater security of a longer agreement. A longer rental period would also encourage tenants to cultivate the soil more carefully, to guard against erosion, and to make improvements to the site. Landowners, however, prefer to keep both their land use options and their ties to landless families more fluid.

Some Pasanos have tried to find land to rent outside Paso, but nearby communities are also crowded. There are some areas beyond the end of the highway, west of Puriscal, where settlement is less dense and land values and rents are lower. Unless a family has a horse, however, the distance to walk to work and the transport of the harvest present problems. As a result, only a few families have tried this area to find better sources of rental land.

A new couple will try to get land from their parents before approaching other people, and there is usually a patrilineal bias in such arrangements. Pasanos say that fathers who have "extra" land that they are not using will feel more obligation to give it to "help out" their sons than their sons-in-law. Technically, a father can rent his land to his married son, but, in fact, I learned of no case of a father charging for land. Instead, fathers "loan" the land without obligation to their sons, or expect only occasional help in their own fields in return. Land is also "loaned" to other relatives, or even, in one case, to a stranger. Whether loaned or rented, the conditions of the arrangement remain tenuous. One married son who had received a plot from his father for several years said, "He gave me land this year, but maybe he won't next year. I have no security; maybe he won't give me land in the future, and I will have to go looking around for other land." In another case, a landholder decided to put into pasture a piece of land he had previously loaned to his son-in-law for tobacco. The son-in-law had to look elsewhere for land to rent. Renters who must pay and those who do not both state, "I am *using* a piece of X's land since he didn't need it this year," and questioning will elicit, "He's charging me ₡200," or, "No, he is *giving* me the land, though he *could* charge me for it."

Table 2.1 shows the relationships between renters and owners and whether a rent is charged for the land. From the point of view of the farmer seeking land to rent, the table shows the number of cases in which he got land from his parents, his brother, his uncle, a relative of his wife's, a distant relative or acquaintance, or a patron. The term "patron" is used for the wealthy landowners who employ landless men regularly and whose employer-employee relationship has been stable over several years' time. Where a landowner can be categorized as two kinds of kin, the case is counted under the closer kin term, but where a patron is also a kinsman, the case is counted under "patron." All kin terms are expressed from the point of view of the farmer seeking land.

Table 2.1 shows a wide variety of relationships between renters and owners. Although the largest single category is that of "distant or no relative" who rents, the table shows that the number of plots loaned is nearly equal to the number rented. As would be expected, most loans of land come from

TABLE 2.1.   RELATIONSHIP OF LANDOWNER TO RENTER,
             BY KIND OF AGREEMENT

Relationship	Loaned	Rented	Total
Parent	6	0	6
Brother	0	1	1
Uncle (FaBr, MoBr)	5	1	6
Affine	4	3	7
Distant or no relative	2	16	18
Patron	4	2	6
	21	23	44

patrikin, but two unrelated owners also allowed land to be used without charge. Patrons use free land as a way of tying steady labor to themselves.

The table confirms the results of conversations with Pasanos and suggests that there is no set pattern as to which kinds of people are approached for land. One man said he only looks to relatives and godparents. Others say they mostly rent from non-kin. Still others say they only ask where they know the household has land it will not be using. The patterns in the table reflect a relatively fluid mode of obtaining land to rent, which leaves a new household with few prescriptions for who can help them and who cannot.

Changes in Land Rentals

Several important trends can be noted briefly in the market for land rentals and in the situation of the landless household in recent years. First, as more and more land is taken out of agriculture for pasture, both the insecurity of tenure and the overall scarcity of land to rent are increasing. The insecurity of tenure, as noted, tends to decrease the tenant's care of the land, and since fallow periods are nonexistent on most farms, such lack of care can only speed the pace of soil depletion and erosion. At the same time, the number of households seeking land to rent is rising, placing even more pressure on soil quality.

Second, the declining yields from rented plots create greater hardships for landless families. One family living in a house badly in need of repair was unable to fix it up because of poor harvests and high rents. "With the money my husband had to pay for rent," said the wife bitterly, "we would have fixed the roof and not get wet when it rains."

Third, the greatest hardships occur when little or no land at all is found to rent. In the year of this research, seven households were unable to find even a whole manzana (.69 hectare or 1.7 acre) to rent, and two families went without land entirely. I was told this was the first time in the history of these households that they had ever been unable to locate at least one small plot. These two households lived off the wages earned in agricultural labor, but wage work is erratic and poorly paid. The diets of all nine families suffered and contained little protein in the form of meat or eggs or beans. With no corn from their own fields, the families often substituted purchased white bread for homemade tortillas, since for the price it would go farther in assuaging hunger. Since tortillas supply a major proportion of protein, calcium, and phosphorus (FAO 1953), however, refined wheat flour is a poor substitute. One woman suffered from swollen hands and feet in the morning and knew that her edema was due to anemia but said, "What can I do? There is no money."

Since dependence on wage labor is so precarious, all the landless families have been pushed to consider other subsistence alternatives. The three main options were all exercised during my year in Paso. One family left to try to find work in the city, and another found a very expensive piece of land to buy. In the third household, one of those unable to find any land to rent, the farmer formally attached himself as a peon to the largest landholder in Paso. His family moved to live on this landholder's farm, and in return for steady labor the farmer received a small plot to work for the following year. Since land purchases are becoming more and more difficult, it can be predicted that out-migration and patron-client ties will be the two options more frequently exercised in the future.

Purchases of Land

The goal of all new households is to purchase land, and, in general, land is bought and sold freely in Paso, without sentimental attachment. The process of sale usually begins with an interested buyer approaching a landholder and inquiring about the possibility of sale. If the landholder wishes to sell, visits will continue until a price is agreed upon. For most of the land bought in Paso, the terms of sale include a down payment and yearly installments until a total sum is reached. When the terms are decided, both parties sign a paper, either a formal legal paper, or, in the past, any paper, stating the terms of the sale, the price, describing the location of the land, and the cosigners, if any. At this point, whether the down payment has exchanged hands or not, the land is considered to be sold. Although a household might fall a year or two behind in the payments schedule, I heard of no cases of default or conflict between the persons involved in a land sale. In several sales, the seller was a woman, but there were no cases in which the buyer was a woman; buyers are usually married men with families.

Land becomes available for sale in four ways. Often, with the distribution of inheritances after the parents' death, some heirs are willing to sell their portion, especially if they have migrated to live elsewhere. Land is also available from time to time in the isolated regions west of Puriscal, and occasionally a Pasano will hear of it and attempt to buy a farm there. Five individuals are absentee owners in this way. In a few cases, sales of land resulted from a decision by absentee owners to sell their holdings in or near Paso. A final way to obtain land is from neighbors or other Pasanos who can be persuaded to sell. One poor family sold a small plot of their land to get some cash; the buyer wanted to expand his adjacent stand of coffee. More often, sales of land between Pasanos occur between family members, sometimes with the desire "to help out" a son or relative. This bias in land sales to kin is one of the few evidences of familism in the community.

Table 2.2 details the kind of kin relationship between buyers and sellers of land for 22 cases. I attempted no systematic survey to record all land sales in the history of Paso, and this table therefore contains only a summary of all recent land sales for which the data are quite complete and also the past cases for which information on both the buyer and the seller is available.

TABLE 2.2.   RELATIONSHIP OF LAND SELLER TO BUYER

Relationship	Sale within last two years	Sale three or more years ago	Total
Parent or sibling	2	4	6
Patrilineal relative (uncle, aunt, cousin, or cousin's husband)	2	2	4
Matrilineal relative	0	1	1
Wife's patrilineal relative	3	0	3
Not kin	6	2	8
	13	9	22

The table shows a strong patrilineal bias in access to land; 10 of the 22 cases (45%) involve the patrilineal relatives of the male household head. In only three cases was land purchased from his wife's relatives. In 36% of the total cases and 46% of recent sales, land was purchased from a stranger, neighbor, or other nonrelative. The table suggests that such sales are rising, but since past data are not complete, such a trend cannot be confirmed.

For the most part, land uses are interchangeable in Paso, and plots are not specialized; all crops can be grown on all the land. The price of land, then, is not affected by its "type"; there are no classes of land according to slope or soil type. Prices are affected, however, by investments such as coffee trees or wire fencing. The particular crop and fertility history of the plot, its location, water resources, and whether it has a title are also important in determining the price.

Land prices have changed greatly in recent years. In the past, the price per manzana fluctuated between ₡80 and ₡600. Six hundred colones was the maximum price before the highway was built to Paso, ten years ago. The price per manzana then shot up to ₡1,500. Today, top prices for a manzana in the community are ₡6,000–7,000, and the mean of sale prices during the period of my fieldwork was ₡3,600 (which includes several larger pieces that were sold at a relatively lower price per manzana). Thus, over the past ten years, land values have gone up as much as 1,000%.

With such a jump in land values, it is not surprising to observe a hopelessness among many landless farmers or small landholders with regard to the possibility of acquiring land in the future. "At these prices, how can I?" is a common statement. In the past, although the prices for agricultural produce were also lower, more moderate land values meant greater possibilities for young families who wished to buy land. This change in the land market is one of the consequences of greater "development" in Paso, stemming in particular from the new road, the electricity line, and the proximity to a good market center and its services. It also reflects the growing population density and the changing land use patterns as discussed for rentals.

A household's economic choices are limited if it does not own land. Government programs to encourage family vegetable gardens have met with little success among the poorest families because they cannot find suitable land on which to grow vegetables. Permanent crops such as coffee and fruit trees are also unwise for the household living on someone else's land. Sometimes these plantings must be abandoned, as in the case of one couple who had lived on the wife's parents' land for 12 years. They had planted coffee, fruit trees, vines, vegetables, and flowers around the house, but left these improvements behind without recompense when they moved to land of their own. The woman explained that her parents would have been angry if they had taken anything away with them: "They were good enough to let us live there, so we owe them all the improvements to the land."

Land Inheritance

The process of land inheritance in Paso involves a number of options, both for the landholder and the heir, which are exercised within a traditional framework of equal rights for all children to their parents' land. At the same time, the patrilineal bias of land rental and sale can also be seen in some cases of inheritance in Paso, as daughters have less claim than sons. A woman's inheritance is added to the pool of the family's land resources; no wife in Paso administered her own land separate from her husband's.

By Costa Rican law, if a husband or a wife dies, the family's total land is to be divided in half. One part is given to the surviving spouse and the other half is split equally among all their children. When the other spouse dies, the remaining half is again divided equally among all the children. If, however, landowners wish to divide their land up before they die, they may do so in any way they choose. "It is completely up to the whim of the parent," testified one son, "but if the land is undivided at death, it must be divided equally." In actuality, land is most often left unsettled until the death of both parents. Though the law says one division should take place at the death of the first spouse, in fact, no such case exists in Paso. Instead, the farm goes into limbo (*está en sucesión*), and children continue using the land without a formal settlement.

When land is to be divided by the heirs after the death of both parents, the customary procedure is a lottery with "cigarettes." As described by one informant:

> We all met on the land to divide it up. There are ten heirs. I went on my wife's behalf and also representing her sister, who lives in Puriscal. Jorge from ——— [another village], who is married to another sister, was also there. Maria sent her son to represent her, but Pedro, my brother-in-law, couldn't come. So nine of the heirs were there. We measured the plot—it's in one piece—with a string 32 *varas* long, and divided it into ten equal parts. The first section had the road going

through it, so we gave that plot another strip farther down. There was also a fine tree on one piece, but we decided to leave the tree where it was and let that person be extra lucky. We numbered the parcels and drove in stakes as markers. Then we made paper lots ["cigarettes"], each with a number on it, and took turns pulling them out of a hat. Everyone was satisfied.

There are a number of cases in which Pasanos divided up the family's lands before their death, and in all these cases, it was the father of the family who made the arrangements, with or without consulting the mother. In such situations, heirs may receive unequal portions, and sons usually receive more than daughters. In one such case, a very wealthy man divided his land in small but equal shares for his daughters, left large pieces for each of his two sons, and left a sizable sum of cash for his wife. In another case, a man with three daughters and one son divided his land so that all received an equal portion, except one daughter whose inheritance was much smaller. She had been the only child of the four supported through high school, and the gift of education was the other part of her inheritance. In a third case, however, the lands were all divided equally among the prospective heirs. As one son explained, "He divided up the land so we could get down to work and plan for it, and so there would be fewer problems when he died." Thus, the landholder's right to distribute land as he chooses reinforces the individualistic thread in Costa Rican life.

One kind of problem with inheritances that are not divided by parents prior to their death is that when the land is in limbo, agricultural decisions are affected. For example, one heir planted a stand of coffee on his father's land, knowing he might lose it in the inheritance lottery. In fact, that plot fell to his sister. Fortunately, her husband did not need the land immediately and allowed his brother-in-law to continue to harvest his coffee while starting a new stand to replace it. Neither party considered swapping their plots, since the coffee stand rightfully belonged to the sister.

While most inheritances are divided amicably by the lottery procedure, problems have also arisen in several instances of inheritance fraud, both recently and in the distant past. In one example, the elderly landowner died very suddenly, and only one son was at his deathbed. The father signed over his deed to the son, with the understanding that the son would divide it equally among his siblings. Instead, the son sold it all to a wealthy storekeeper from Puriscal and went to another part of Costa Rica with the money. The remaining heirs had to migrate elsewhere, too, because the storekeeper turned the whole property into pasture and the heirs had no land to work. Several similar cases were recounted, but in none of them did the disenfranchised heirs take any action against their brother.

Land registry, with a legal deed, is the means by which deceit of this type usually takes place. A deed must be obtained after a formal survey of the land is completed and then the ownership must be registered at the canton level. Not only does a formal deed protect an owner and increase the land's resale value, it also can serve as collateral for loans. The cost of surveys and legal fees can be enormous, however, and several disputes have arisen recently in which one son paid all the costs of the deed and registry and, to repay himself, took an unequal share of the land. There is no tradition to handle this new aspect of inheritance, and both sides have remained resentful for years. Once a deed is made, the only recourse for disenfranchised heirs is an expensive legal proceeding.

With land registration comes vulnerability to taxation. At present, the Costa Rican government levies no taxes on any landholdings under ₡10,000 in value, and owners may declare the value of their land as they wish. Few Pasanos pay any land taxes, and those who do pay only a small sum. Taxation rates are substantial only for very large farms of a size not found in Paso. Some of the larger landholders register their land under the names of several family members and thereby avoid most of the taxes they would otherwise pay. Although many farmers cite fear of taxation as the cause of their reluctance to register the land, when asked what they would do if deed and registration were free (an AID program in another area of Costa Rica pays these costs) individuals replied with alacrity that they would register their land to keep it safe: "It would be better to pay taxes and have a deed." The value of land registration is seen, therefore, as outweighing the risk of possible taxation.

Heirs who receive land while their own families are small or still growing usually put the land to work immediately. In a few cases, the heirs are grandparents whose need for the land is much less than that of their married children. In such cases, heirs have a variety of options, but in fact often tend to leave the land in the family. One widow who had migrated to the capital city received land in Paso at the death of her father. Her eldest son still lived in Paso and offered to buy the land from her. She agreed and wanted to sell him the plot at a token price, since none of her other children wanted it. The son urged her to consider her other options: she could charge a higher price and keep all the money for herself to take a trip or buy a house; she could also rent the land and use the income in various ways. Although the woman remained firm in her decision to sell cheaply to her son, his care to point out more equitable distributions can be seen as an attempt to avoid future conflicts with his brothers and sisters. In another case, a married woman sold her inheritance to her son and built a fancy new house for herself and her husband. In a third situation, the heir rented out his land to a landless neighbor. Thus, the landholder's right to dispose of land as he or she wishes is upheld in practice, but there is a clear tendency for relatives to have easier access than strangers to land for sale.

ACCESS TO LABOR

Agricultural work in Paso is carried out primarily by the male household head, aided at times by his wife and children. A few childless couples illustrate that children's labor is not necessary to maintain an average standard of living; the work of one man is adequate for most seasons of the year. Many households do use their children for agricultural tasks, and many also hire agricultural laborers.

Household Labor

Most Pasanos state that the husband is the "head of the household," although many also caution that the couple must work together, "like a yoke of oxen." As with oxen, if one side pulls ahead or lags behind, the team comes to a stop. These conflicting notions of marital relations carry over into the allocation of household labor. In general, there is a division of labor in which women are in charge of domestic duties (cooking, dishes, cleaning, laundry, and child care) while men are responsible for running the farm or bringing home wages. The husband is recognized, however, to have control over family labor, and he can decide, technically without consulting his wife, where she and any children will work during the day. I observed many situations in which a wife deferred to her husband's authority, since he had the best grasp of the overall needs of the farm and the urgent tasks awaiting with certain crops. "I need Zinia to help me with tamales," said one woman "if my husband doesn't need her to pick coffee." At the same time, the yearly round of tasks is familiar to girls as well as to boys, and the periods when children's labor is needed outside the home are anticipated by the women even before their husbands may discuss it. Most agricultural tasks are done by men, but women and children help at key points such as during the coffee harvest (the division of labor by sex will be discussed in more detail in Chapter 4). Once sons have finished the six required years of schooling, they usually work full time with their fathers.

While a son is unmarried and living at home, he "works for the family." This phrase means his labor is at the disposal of his father, without pay, for work on the family's crops, to do errands, or to work with his mother. Sons are seldom sent by their fathers to do wage labor for other farmers, though their fathers have the "right" to do so. Some young men arrange by themselves to do wage labor for another household, if they can find work and if their fathers do not need them. In these situations, their wages are usually, but not always, turned over to their parents to help in household expenses.

Sons learn agricultural methods by working with their fathers in the fields, but some parents feel it is important for sons to work small plots by themselves, "to learn." Where families have extra land, the son can use a parcel near home; other families permit their sons to rent land. In both such cases, the son retains the proceeds and can spend money as he pleases or save it toward marriage.

Some parents do not allow their sons to develop independent finances, a situation that has repercussions when they decide to marry. One woman explained:

> My father kept my brothers working in the house like servants and so when they married, he had to pay for everything—house, dishes, everything. Carlos [a neighbor] doesn't do it that way. From the time his sons are adolescents, he gives them a pig to raise and sell, crops of their own, always letting them do something themselves. So when they decide to marry, the sons have their own money and pay for it all themselves. Last month, when his son married, Carlos paid only for the fiesta. I think that is a better way.

The choice of whether a son is free to work independently seems to lie with the father, although the personality of the sons is involved too. There is no generally preferred rule, and both patterns of father-son relations are widely chosen. After marriage, most sons no longer work for their fathers without pay, and fathers feel no hesitation about hiring their sons as peons. Thus, if sons are not allowed to become economically independent before marriage, this process is completed when the new household is established.

Hired Labor

Over half of the agriculturalists in Paso hire labor at some point in the agricultural cycle. Some hire help only a few days a year, to help with the hardest part of tobacco terracing or with the pressure of ripening coffee that must be picked or lost. Other farmers hire peons regularly and employ workers more than 300 days per year. Usually, landed farmers employ landless ones, but the reverse also occurs, and there is much employment within these groups as well. In Paso, any man who works for another calls himself a peon, though the next day, he and his employer may reverse positions. (*Peon, employee,* and *worker* are used here interchangeably.)

Wage labor patterns in Paso are flexible and fluid. Both employers and employees maintain many ties, since at different times of the year each experiences scarcity. Since not all peons work equally well, employers hire different peons at different times and try to avoid dependence on one worker. By having hired several different workers, the employer maximizes his chances of obtaining help in seasons of peak labor demand. Likewise, at periods of low demand for laborers, the peon who has worked for many people has more places to inquire for work. Having ties to several employers also gives the employee some choice in his working conditions and his boss. According to the law, an employer who hires a peon on a fixed basis, be it six days a week every week, or one day a month, must pay minimum wage and social security. This law further encourages the fluid labor market, as employers avoid being liable for these costs.

Both employers and employees refer to wage labor as "working with" someone or "helping" them, although it is understood that such help is always paid. The unmarried son of a middle-sized farmer mentioned that half of his tobacco terracing was done by a neighboring middle-sized land-holder, a man who had also terraced a large tobacco plot of his own: "He did it out of friendship." Was he paid? "*Of course!!*" Workers all begin at 6 A.M. and finish at noon, since most agricultural work is done in the rainy season, and the afternoons are rainy. When the afternoons are clear, however, a peon may be asked to continue working, for which he will be paid accordingly, or he may go home and work on his own fields in the afternoon.

To arrange for a peon to work for him, the farmer visits the peon's house himself, or sends his children, and asks the peon if he is free to work on a certain day. If the peon is available, the employer will tell him in which fields they will work and what tools to bring and will meet him in the designated field in the morning. All Paso employers work in the fields, usually beside their employees, and are generally expected to work harder and longer than the peons. After that day's work, if the employer needs more help, he will ask the man if he can come back the next day, or the following week.

A man who wishes to do some wage labor can also visit the home of a prospective employer to ask if he has any work for the following day (he would not send his children to ask). At the end of a working day, a worker may ask his employer if he should come back again the next day or the next week, if he is needed. At certain times of the year, however, it may be hard for the peon to find work, and he may have to ask at many different houses, hence the value to him of a network of potential employers. Although landless households depend heavily on wage labor, they will refuse to work for others at certain times when their own crops on rented land need attention. As one woman said, "My husband won't offer to work for Tomás these days because our corn needs weeding and Tomás would grab him for the whole week if he could."

Payment customs are flexible in Paso, but the wage is fixed. The agricultural wage in Paso is ₡6 a day (one colon equals $.12U.S.) for a six-hour workday for all tasks except tobacco terracing, which is paid at ₡10 for six hours because the work is so strenuous. Most employers are willing to pay peons at the end of each day if so requested. Peons state that they will ask for payment at that time if they need cash but otherwise prefer to wait until Saturday for a lump payment for the week or fraction that they worked. No wage labor is done on Sundays. Some employers offer wages wholly or partly in kind: corn, beans, coffee, and sugar. Such payment benefits the employer by assuring him a set price for his produce but also helps the employee because the foodstuffs are cheaper from employers than they would be at the stores. Some peons may ask a patron for a cash loan and then work for him the number of days necessary to pay it off. Others obtain grain in this same way, then work for the employer until its value has been returned. In all these forms of payment, the employee decides his preference and he is neither cheated nor the value of his wages discounted by the option to be paid in kind. "We're not stupid!" exclaimed one poor woman when I probed for more details of the value of wages paid in kind.

Wages in Paso are below the legal minimum of ₡2 per hour. One might suggest that this is due to excess supply of labor in the rural areas, and there is some evidence to support this position. Both large landholders and landless workers, however, have said that if the minimum wage were enforced, there would be less wage work available, since employers cannot afford to pay more than they are now paying. In addition, the fact that landless families also hire people at times may contribute to their acquiescence to these sublegal wages.

There are clear obligations between employers and employees, though Pasanos rarely talk about them. The peon is expected to work on a task even when his employer is not around, and some employers complain that peons do not work so hard if they are alone. One large landholder gave this as his reason for not putting more of his land into grains to sell. In addition to fair payment, the employer is expected to be concerned to some extent with the welfare of his workers. One boy cut himself badly while working for a man, and, although the boy did not work for him regularly, the employer felt it was his duty to take him into his house and care for him until he was better. In another case, a man who was injured refused to work again for that employer because the latter had not come to visit him at home or even ask about his injury. There is also one case of political loyalty from a Pasano who worked regularly for a man from a neighboring community. The peon voted against his own political preference in order to support the political party of his "patron." In Paso, however, the political power of employers is minimal, since the community is relatively homogeneous in politics. In fact, the entire Puriscal area is famous for its allegiance to one political party.

Changes in Hired Labor Availability

Employers in Paso complain that peons are becoming scarcer in the peak labor times, especially August and September. Some say that more people are leaving Paso in recent years, but rates of out-migration do not uphold this position. Other landholders attribute the labor shortage to the attention landless farmers are paying to their own crops. This explanation seems inconsistent with the decline in farm size in Paso, which tends to increase the number of persons looking for wage labor. Also, the proportion of landless who work rented land has decreased slightly in recent years, not increased.

While it would be difficult to prove whether there really is, to any measurable degree, an increasing shortage of peons, the land use shifts discussed in Chapter 1 suggest reasons why it might seem so. On the one hand, land in coffee is increasing, and thus landholders who have increased their plantings now need more harvesters in August and September. The coffee harvest overlaps with the period of tobacco terracing, when even small tobacco farmers want help in the difficult work of ridging up the soil. Since tobacco plantings by landless families have also increased greatly in Paso, this trend pulls their labor onto their own fields precisely at the period of increasing demand from employers. Hence it is not surprising that there is much discussion of a labor shortage in August and September, although this shortage is not noticed earlier in the year. As land for renting continues to become scarcer, this trend will probably reverse. It will also be interesting to note if, in the meantime, wages rise during parts of the year as employers compete for employees.

SUMMARY

The life cycle of the Pasano household and the patterns by which it obtains land and labor reflect all three of the themes discussed in Chapter 1. The cooperative, nonfamilistic thread is seen in the willingness of some households to give land to rent or even for sale to strangers, neighbors, or other non-kin. Counteracting this pattern is a clear bias in both the sale and rental of land toward "helping out" kin first, especially patrikin. There is a fluid, atomistic market of wage labor ties, as the majority of those who work as peons regularly are not tied to one specific family but instead maintain relations with a number of different families. A similarly fluid market of land rentals prevails.

The increasing scarcity of land is pushing Paso toward a more rigid social order in which those who will not inherit land from their parents must remain landless because they cannot afford to buy land at the soaring prices in the area. Out-migration and closer patron-client ties are two responses to this situation. Population pressure from the landless has repercussions as well on the way rented lands are used, further adding to soil erosion problems.

3
Consumption Patterns, Social Interaction, and the Five Strata

Recent research has emphasized the importance of internal differentiation within rural communities (Cancian 1972, 1979; Dewalt 1979a; Pelto and Pelto 1973). Particularly in the analysis of agricultural change and farm decisions, different resource mixes affect how farmers choose production options (Hildebrand 1977). This chapter will explore the different standards of living of wealthy and poor Pasanos, the patterns of their social interactions, and recent changes in political power and leadership. The strata defined here will form the basis for the analysis of agricultural change in succeeding chapters.

HISTORICAL DEVELOPMENT OF STRATIFICATION

Some people have nice houses and some don't—some have bigger ones and some small, some newer and some older, but we are not all the same and we can't be. People will always be different and it's better that way. [A Pasano farmer]

. . . The stratified society is distinguished by the differential relationships between the members of the society and its subsistence means—some of the members of the society have unimpeded access to its strategic resources while others have various impediments in their access to the same fundamental resources. [Fried 1960:721]

As in most rural societies, access to land and its productive potential is unevenly distributed in Paso. From the days of Paso's earliest settlers, some people owned land and others did not. The original five families who came to Paso from the central valley claimed huge tracts of land for themselves. No records remain to help us determine if there were any smaller farms among these five, but informants do not remember any such farmers or farms. By the time the children of the original settlers were adults, however, there were some families who lived primarily or entirely by wage labor, working as peons for the owners. Forty percent of the landless families in Paso today are descendants of one man, son of one of the original founders. This peon was tied closely to his brother, who went on to become the largest landholder in the community. The descendants of the landless brother are among the poorest, least well fed, and least respected members of the community, but they are first cousins to the children of the wealthy brother, who have for the most part remained substantial landholders.

The history of land tenure in Paso seems to be quite different from the sequence described by Sandner (1960) for "Turrubares," nearby. The area Sandner labels as Turrubares includes parts of the canton of Puriscal and is no less isolated than Paso, though it is somewhat lower in altitude. Sandner describes the earliest settlers as slash-and-burn farmers who used the land briefly, then moved west. Their farms were purchased by investors from the capital city, who Sandner feels pushed out the frontier settlers by driving up prices. These investors became absentee cattle ranchers and sometimes attached the former farmers to their haciendas as peons (Sandner 1960:56).

Such haciendas do not characterize Paso and its neighbors, and the area seems to have followed a different sequence. While there have always been a few absentee-owned cattle ranches, usually at the edges of the community, they have not displaced the peasant farmers, nor strongly affected the character of community life. Pasanos have seen many outsiders come and go, buying and selling different plots of land, with little social or economic impact on the permanent peasant community.

Margolies (1977) suggests a process similar to Sandner's for frontier areas in general: wasteful, shortsighted agricultural methods are adopted by frontier settlers, and when soil fertility drops they are pushed out and land and wealth become concentrated in a few hands. The resulting polarization of the community bears many similarities to the processes in Paso today: landless families find it nearly impossible to buy land, small farms suffer greatly from the drop in soil fertility, and much of the land of the community is owned by a few families. Several aspects of Margolies's sequence do not apply, however. There is no evidence to prove that early Pasanos were using "wasteful" methods of land use, as discussed in Chapter 1. Although the smaller farmers do feel squeezed, they have responded with new agricultural techniques and have not yet been forced to leave the community. Out-migration has been mainly from landless families and young people, not established small-farm households. Finally, as is clear from the brief history of Paso, land concentration is not new and cannot be said to be purely a response to soil depletion.

Several aspects of life in Paso prior to the last 20 years tended to modify the effects of the unequal distribution of land. First, landowners rented land freely, and landless people talk of the scarcity of land to rent as a new and unjust problem. Second, the overall standard of living in Paso was based much less on purchased goods than it is today, and the gap between the consumption patterns of the wealthiest and the poorest was not as great. Some wealthy landholders had large houses, kept fine horses, and ate well, but they did not wear shoes, and their clothes as well as their household furnishings were mostly homemade. Furthermore, land was available for sale, and renters could reasonably hope to become owners. The impediments to access to resources, then, were surmountable, and fixed social classes did not evolve in Paso. Only recently, with rapid population growth and the advent of an export market for beef, has the private ownership of land come to mean a true impediment in access to resources for a portion of the community. Patterns of social interaction still reflect the egalitarian traditions of the past, but firm distinctions of wealth and status are beginning to appear today.

THE FIVE STRATA

During the year of this research, it was evident that Pasanos make different agricultural decisons according to the amount of land available to them. Further analysis revealed that neither family size nor kinship ties diminishes the primary importance of access to land. In addition, many aspects of a family's standard of living depend on the amount of land it owns. Therefore, I chose land ownership from among the many variables possible to delineate stratification in Paso. It should also be noted here that all Pasano households, with the exception of the economically inactive elderly, are made up of farmers, and they depend on land to make their living. There are two storekeepers, but they both farm in addition to tending their stores. The schoolteachers and other specialists who serve Paso live outside the community.

I have distinguished five strata in Paso. The concept of *stratum* is used here rather than *social class* because these groups are fluid and the lines between them somewhat arbitrary. Although Pasanos make similar distinctions among households in terms of landownership, housing, dress, and agricultural decisions, they would not necessarily divide the community into five groups. These groups do reflect differential access to resources, however, and thus are called strata to emphasize their basic inequality.

Pasano households can first be divided into land and landless categories. The families in the landless category are quite a diverse group, and it is necessary to separate those households that will someday inherit land from their parents (labelled here the "heirs") and those who are children of landless parents (labelled the "landless"). Both groups have difficulty renting and buying land, but heirs can often get land from their relatives and can plan for a future in which they own land; the landless cannot. The agricultural decisions as well as the consumption standards of these two groups differ somewhat. There are 17 landless households in Paso and eight heirs.

Based on a subjective assessment of standard of living, the landed families are divided into three groups: "small," "medium," and "large." The line between "small" and "medium" is drawn somewhat arbitrarily at 7.0 manzanas. The average farm size in the community is 14 manzanas, but this figure is skewed by five large farmers who own over 75 manzanas. When these five are omitted, the average falls at 7. The smallest farm is .25 manzanas, and the group owning from .25 to 7.0 (25 households) is more homogeneous than when the line is drawn at 14 manzanas. There are no families with farms between 28 and 49 manzanas, and therefore the "medium" stratum contains farms from 7.1 to 28 manzanas (17 households), and the "large" stratum includes all those over 49 (8 households).

Table 3.1 shows that just over one-third of the community owns no land; two-thirds of these will not inherit land, and one-third will. Small landholders make up 33% of the population, and the medium and large landowning households make up the remaining third. Land concentration is pronounced; the large farmers with 11% of the population own 67% of the land in Paso. The small farmers who make up one-third of the community own only 7% of the land.

Differences in Standard of Living

The strata vary on a number of parameters, but four aspects of standard of living will be discussed here: housing, furniture, dress, and diet.

TABLE 3.1. FIVE STRATA IN PASO

Stratum	Land owned, in manzanas	Number of households	Percentage of population	Percentage of all land owned
Landless	0	17	24	0
Heirs	0	8	11	0
Small owners	.25–7.0	25	33	7
Medium owners	7.25–28	17	22	26
Large owners	49–167	8	11	67
		75	101	100

Three types of housing are found in the community. These types represent both the levels of wealth in the community today and also an evolution of housing forms over the last hundred years. Type I houses are one- or two-room palm-thatched houses with walls of vertical poles lashed together with rope. Older informants say all but the wealthiest families used to live in these Type I homes, which are similar to the former Indian dwellings of this area (Lothrop 1926:33). A fortunate few in the past had Type II homes of planed boards, nailed horizontally, with a metal or tile roof. Now, the majority of the community lives in houses such as these; they usually have a dirt floor in the kitchen, a board floor for the other rooms (two or more), and a roof extending out beyond the front of the house to form a dirt-floored porch. In recent years, a third type of house has become popular, built by carpenters who come in from outside Paso. Type III houses are made of brightly-painted purchased lumber, and all have metal roofs. The floors are varnished wood, cement, or occasionally tile, and the house usually has glass windows. The arrangement of the rooms is distinctive, with a small tiled porch in the front of the house and a guest room opening off it. Such houses are considered complete only with curtains and manufactured furnishings, and they are virtually indistinguishable from housing in many urban areas of Costa Rica. Nearly a third of the community now lives in this newest type of house.

The strata are dispersed as would be expected among these three house types. Type III, painted, glass-windowed, urban-style houses predominate among the large landholders. Medium landholders are distributed half and half in Type III and Type II homes, while small landholders, heirs, and landless families live predominantly in the Type II houses. Type I houses, with palm-thatched roofs and walls of lashed poles, are found only among landless households.

Wolf, in discussing social relations in the open peasant community, indicates that one of the most important types of status behavior is "the ostentatious exhibition of commodities purchased with money" (Wolf 1955:519). In Paso, furniture and clothing are especially important in this respect. The Type III house is not complete without a matched, stuffed, vinyl-covered sofa-and-chairs set with a plastic-topped coffee table. A plastic-topped aluminum table with matching side chairs is the preferred dining room furniture. Once electricity was brought into Paso, several wealthy families bought stoves, refrigerators, and televisions. Radios used to be items of status, but now almost all households own one.

Ordinary households own a homemade wooden table to eat and cook on and wooden benches or stools to sit on. Chairs with backs are quite rare, but benches with backs made of wooden slats are commonly found. For many families, beds are raised wooden platforms, with several blankets for covers.

Wealthier families have mattresses, sheets, and pillows; the wealthiest households have purchased metal bedsteads or even Formica headboards with shelves. The poorest family in Paso has no benches at all to sit on—only the bed, which takes up most of the tiny one-room house. Their kitchen lean-to has the typical stove made of a raised wooden platform filled with packed earth and ash. This stove type is similar for all strata, varying only in size and whether or not it has a chimney. All Pasanos cook with wood, though some families own kerosene or electric stoves as well.

High-status clothing for women includes pants suits and heeled shoes. Most women, however, wear simple A-line cotton dresses and plastic shoes or sandals without socks. Long-sleeved shirts are a mark of affluence for men, since agricultural work is always done in short-sleeved shirts. Men usually wear rubber boots or synthetic leather shoes, but wealthier men may boast shoes of real leather. A few of the poorest Pasanos may occasionally go shoeless while working, but the vast majority wear some sort of shoes at all times. Men also generally wear socks. Purchased clothing, as opposed to clothing made in the community, is always a mark of greater wealth, for both sexes. Women's handbags are also an opportunity for the display of wealth.

The traditional Paso diet is made up of boiled rice and beans eaten together with tortillas. Seasonings are bland—salt, *culantro* (coriander), and garlic are the common spices. Coffee and *agua dulce* (hot water with brown sugar) are the main beverages, though families with cows usually have milk as well. Some squash is eaten at certain periods of the year, but the Mesoamerican triad of corn, beans, and squash has been altered by the addition of white rice, the major bulk of the meal. Eggs, vegetables such as cabbage and tomatoes, and some onions, potatoes, and noodles are occasionally eaten as well. Meat used to be a market-day addition for many families but now is rarely purchased; the price has gone up so high that most families now eat meat only for special occasions. The distinctions between "hot" and "cold" characteristics of food are observed in Paso, as in so many Latin American countries.

The differences in diet among the strata come mostly from the frequency with which certain items appear. Wealthier families have meat several times a week and may also have cheese and fish on occasion. Cookies and crackers, as well as canned fish and fruit juices, are found often in the well-to-do homes but are not unknown to poorer Pasanos. Some children are given money to buy candy at school recess every day, while other children get candy only on special occasions. Families experiencing hardships may go without beans or tortillas for a period, and one landless woman said she had gone without meat for so long that her stomach could no longer tolerate it. Large quantities of sugar are consumed daily by Pasanos in their coffee, *agua dulce*, and milk. Refined white sugar used to be a mark of status but is increasingly replacing the traditional hard cakes of brown sugar, as the price of the latter rises. Manufactured foods such as bread, noodles, and soft drinks are all more common in the diet now that the new highway makes their delivery easier. Such purchased items are rarer, of course, in the diet of the less affluent families.

These differences in standard of living and in consumption patterns are a subject of considerable interest to Pasanos. The latest purchases of local families are quietly discussed by the rest of the community. Thus, the rising standard of living of the wealthy is observed by all strata and sharpens distinctions that used to be hazy. This conspicuous consumption does not affect Pasanos' basically "Puritan" attitudes, however; all groups continue to exhibit a "high evaluation and practice of manual labor, ambition, and frugality" (Goldkind 1961:38).

Fluidity and Stability of the Strata

Two indices can be used to explore how fluid or stable are the social strata just defined. Given the process of land acquisition defined in Chapter 2, there emerges a pattern of increasing farm size over the lifetime of the household. Table 3.2 shows the mean length of marriage by stratum. Households of widows and unmarried siblings are omitted. These figures show that large landholders tend to be older and married longer than all the other groups. Heirs, predictably, are younger than any other group, and the small and medium strata form a continuum of increasing farm size and increasing length of marriage. Landless families are also young, in comparison with the landed strata, but not so young as the heirs. The existence of a lifetime pattern of increasing farm size can be tested with a contingency coefficient. The relationship between years of marriage and land owned ($C = .36$) is significant at the .05 level (maximum value for C in this calculation could be .816). This figure shows that the correlation between the age of the family and its land accumulation is greater than one would expect from chance, but it is still lower than the .816 that would indicate completely fluid strata. While continuous land accumulation is the ideal of all households, this figure illustrates that mobility is somewhat blocked.

TABLE 3.2.   MEAN LENGTH OF MARRIAGE BY STRATUM

Stratum	Years married
Landless	15.6
Heirs	8.9
Small	19.1
Medium	23.2
Large	32.6
$C = .36$	

A second indicator of stratification is the extent to which the strata intermarry (Cancian 1965). Since the categorization of Pasanos into five strata makes the numbers involved quite small, Table 3.3 explores the randomness of spouse choice between all living couples in Paso whose parents have land and those whose parents do not. The Chi-square test shows p is less than 0.1 but more than .05, thus suggesting there is some evidence for nonrandom marriage among the landed and landless families, but the pattern is not strong.

TABLE 3.3.   INTERMARRIAGE AMONG LANDED AND LANDLESS STRATA

Parents of couple	Observed frequency	Expected frequency
Landed/landed	33	29.4
Landed/landless	18	25.2
Landless/landless	9	5.4
	$.05 < p < .10$	

Some of the variance in the observed and expected figures in Table 3.3, however, may be due to the fact that there is not an equal number of males and females in each of the landed and landless categories. Also, it distorts the data to lump together all living couples, regardless of age. It cannot be expected that a 50-year-old landed woman is as likely to marry a 20-year-old landless man as a 55-year-old landed man. Thus, Table 3.4 explores whether or not spouse choice is random within two age cohorts. The current Pasano population is divided into younger marriages, between partners aged 20–40 years, and older marriages, between partners aged 41 and over. Seven cases that span the two cohorts are omitted. The older cohort shows that spouse choice is random in regard to the landedness of parents. In the younger cohort, there is slight evidence that the stratum of one's parents affects marriage choice, but the significance level is below that of the whole sample in Table 3.3. Additional tests to see if the younger cohort's nonrandom marriages tend to occur more among landed males/females or landless males/females indicate that there is no significant difference by sex in the patterns of spouse choice.

It seems clear that much of the appearance of strata endogamy in Table 3.3 is primarily due to the incorrect lumping of age cohorts. To the extent that there is some slight tendency toward strata endogamy among the younger couples, this finding tends to support the general trend toward increasing rigidity of the lines between the landless and landed in Paso. Thus, although wealth and access to land are factors assessed by Pasanos when they choose spouses, other considerations are also important, and there is no barrier to intermarriage between the children of landed and

TABLE 3.4. INTERMARRIAGE BY AGE COHORTS

	OLDER COHORT		YOUNGER COHORT	
Parents of couple	Observed	Expected	Observed	Expected
Landed/landed	13	12.1	15	12.9
Landed/landless	11	12.7	6	10.2
Landless/landless	4	3.2	4	1.9
	(not significant)		$.10 < p < .15$	

landless couples. In fact, four of the six such "mixed" marriages are between children of landless and large landholders.

Pasanos would not be surprised at these figures. While they recognize that some people do marry for self-interest, they stress the importance of love in marriage. When there is no love between the couple, marriage is seen as *muy duro* (very oppressive or difficult), and love is generally recognized to be quite capricious when faced by such mundane matters as inheritance. This emphasis on the personality and character of one's prospective spouse fits the egalitarian theme of Pasano life and also reflects the more fluid economic situation of the recent past. When families without land could reasonably expect to acquire it with hard work, the choice of spouse could be somewhat distant from an assessment of family assets. Both the pattern of increasing farm size over the course of the marriage and the random selection of spouses illustrate the historical strength of egalitarianism in the community.

INTERACTIONS AMONG THE STRATA

The patterns of interaction among the strata in Paso can be divided into three parts: social, economic, and political.

Social interactions in Paso vary quite sharply from those described by Goldkind for a rural community in the central highlands of Costa Rica (Goldkind 1961). Goldkind distinguishes two major social groupings based on access to land. The landowners and commercial tenants form one group; members of this group see each other as equals and feel themselves to be the "community." These people "carry on their economic enterprises with some degree of satisfaction, aspire to better their economic situation, and have at least a possibility of doing so and accumulating some wealth." They organize the yearly round of religious services, school activities, fiestas, and other community events. They discuss national and international politics and sports, read newspapers, own or listen to radios, and play soccer together (Goldkind 1961:375).

The second group consists of the landless laborers and subsistence tenants, who "are not regarded by members of the dominant status group as personalities of equal human dignity" (Goldkind 1961:375). Goldkind points out that they may exhibit patterns of deference, acting "almost as though they were mentally retarded" when with "their betters" (Goldkind 1961:376). The quality of their interactions among themselves is quite different. These poorer households may make up as much as one-third or more of the population, but they struggle for bare necessities and are often malnourished. Recent competition over land, the rising prices of land, and low wages give this group little hope for future betterment (Goldkind 1961:375).

These generalizations apply only in part to Paso. The satisfactions and aspirations of heirs and landed Pasanos are similar to those noted by Goldkind, and the landless stratum does not share them. In Paso, however, a number of landless households engage in the "yearly round" of activities that Goldkind describes. The number of landless who do not participate in community events is greater than for other strata, and they report lack of appropriate clothing and money as their reasons for nonparticipation. At such festivities, however, all the strata chat and interact freely. Although some landless men were observed to be more quiet and reserved among landed men than they might be among their own families, others joked and declaimed on current issues as equals. Men who worked regularly as peons were the most likely to show signs of status differences, but this behavior occurred primarily in the work context. The more exaggerated patterns of deference Goldkind describes were not observed in Paso, nor did the landless form a clearly defined and homogeneous group in social interactions.

The contrast between Goldkind's observations and the situation in Paso is revealing. Goldkind's work was done in an area where population densities and the concentration of land have been evident longer. The land shortage there and the greater proportion of wage laborers may account for the development of such elaborated deference and distance among the strata. Paso's poorer strata are only beginning to face the behavioral implications of obtaining all one's livelihood directly from wage labor. Likewise, their hopelessness of someday becoming upwardly mobile is newer, and for many, their pessimism has not yet reached the point of accepting an irrevocable gap between themselves and the landed community members.

Another context of social interaction is in the *pulpería* (general store), where men and boys gather in the late afternoons and evenings. Such occasions illustrate the egalitarian ethos in Paso. Regular visitors to the store tend to be the people who live close by, but they span a wide range of ages and strata. Women may come to the store to make purchases, but do not stay inside to talk. The group gathered in the evening may number as few as three or four or as many as fifteen on a particularly social night. Weekend evenings attract more men and boys, and the street outside the store may hold four or five knots of men, chatting or playing cards. The men drink beer, rum, or soft drinks, but unlike the interactions in a bar, the primary occupation there is not the alcohol. Discussions of the weather, the latest community meeting, national politics, and sports dominate the scene, and the participants do not manifest any common patterns of deference or greater respect for wealthier Pasanos. Although one man's success in certain ventures or differences in family resources may be discussed on certain occasions, it is implied that the underlying value of each person is separate from his or her worldly circumstances.

Aside from community functions and visits to the *pulpería*, the major form of social interaction in Paso is *paseando* (going visiting). "Going visiting" implies a rather formal occasion—the visitor usually bathes and puts on fresh clothes—but the visit is ordinarily not planned or announced in advance. Whole families can visit, or parts of families, or individuals. "Being visited" requires hospitality, the offering of some refreshment and perhaps also a meal. When members of a family get together, they may simply sit quietly, but more often they talk, catching up on news. Visiting is not necessarily a gay or convivial occupation. It was not rare to see a visitor sitting alone on a porch or lounging near a house. If the visitor were given some food, the required hospitality would have been extended, and no further contact was necessary. The restful, nonconversational kind of visit seemed to provide a respite for some Pasanos from their own unhappy or hectic home situation. For poorer Pasanos, the food given at some wealthy homes was also the object of visiting.

Visiting can take place between neighbors, kin, or relative strangers, though the latter occurs most often when the visitor has some business to transact with the host family. Obviously, this range of people who commonly visit means that this form of interaction crosses the boundaries of the five strata. Since many kin are in different stages of the farm cycle, visiting even close kin may mean visiting people of quite different standards of living. There seems to be some tendency for less wealthy households to visit the more wealthy, though the reverse also occurs. In one case, one sister from a small landholding family married a landless man and remains very poor, while her sister married a well-to-do widower and now lives relatively affluently as the wife of a medium landholder. The children of the poorer sister occasionally visit their aunt and cousins in the wealthier household, though they seem somewhat awkward in doing so, and return visits apparently do not occur.

In some contexts, kinship modifies the distance between the wealthy and the poor. The largest farmer in Paso, who is generally recognized to be the wealthiest as well, is first cousin to the poorest of the landless farmers, and they converse as equals in *turnos* and other community contexts. In other situations, their relationship is more distant, especially when the poorer man works as a peon for the other. There seems to be no concern to equalize the positions of kin.

The social interactions of "going visiting" should not be exaggerated in importance; many households have relatively little contact with others. Some Pasanos assert they have neither gone visiting nor been visited in years. Some families say simply, "We don't go visiting." Households in close physical proximity to other kin often visit more informally, dropping in to talk casually during the day's work. Such sociability is more common where a woman and her daughters live near each other, perhaps because the women are around during the day. Some houses that are close together, however, never exchange visits. In some cases, this lack of contact reflects animosity or conflicts, but more commonly it reflects the atomistic self-sufficiency of the nuclear family in Paso. Frequent contact with friends and relatives is simply not an expected part of adult life, for most Pasanos.

This atomistic pattern means that news and gossip travel erratically. The dispersed settlement pattern combined with the physical isolation of many houses that are surrounded by their fields leaves many adults with little opportunity for contact with others. The husband's weekly visit to town is, for some, a chance to catch up on the news, but other men go and return having conversed briefly, if at all, with others. Students who attend school are sometimes a source of information, but I was often surprised that events of considerable importance were often unknown across town, even in the homes of kin. At times, adults seemed to be trying consciously to avoid talking about others, perhaps following the lead of the priests who regularly spoke out against gossip. For example, the only serious crime in the community for many years occurred during this research, when the home of a well-liked family who had lived in Paso for several years was robbed while they were at mass. A young man from a landless family was accused and was sent off to jail in the capital city. Many families I talked with knew only vaguely about the incident and had no particular interest in it. Most people who discussed the case at all felt the boy's guilt was not completely established, but they had no idea who had done it. The overall tone of these discussions was one of disinterest, except in the case of the boy's mother, who passionately defended him.

Economic and Political Interactions among the Strata

The primary form of economic interaction among the five strata is agricultural employment. (Informal borrowing and cosigning for loans will be discussed in Chapter 8.) The range of relations between peons and their employers is wide: some workers are hired only sporadically, others regularly but with different employers, and others are tied to working regularly for one farmer.

The five strata are all involved in wage labor. Of the 15 economically active landless families, roughly half maintain ties with several employers; the rest work almost exclusively for one or another of the large landholders. There is also one heir who works regularly for one employer. Small landowners and their sons may occasionally work for others, but medium landholders rarely do. The sons of both medium and large landholders will hire themselves out to friends and neighbors as wage laborers, however.

Landless Pasanos who are tied as peons to one employer approach having the kind of patron-client relationship common to many peasant communities. The economic contract between patron and client (Johnson 1971a:115) includes "extras" for both sides that are not normally included in more fluid employer-employee relations in Paso. Peons not only expect to be hired much of the year, but they also usually receive a small rented plot, and sometimes a house site, building materials, and help in emergencies. Patrons are not obligated to give work when they do not need help, but in fact they rarely turn down a request for grains or money to be repaid in labor. In return, the peons may be called on to do unpaid errands and odd jobs and also give up the benefits of having several employers.

Several aspects of this situation in Paso are unlike many patron-client relations discussed for other communities. Patrons in Paso were never observed to act as intermediaries with government organizations, either by protecting peons or by helping them gain access to government programs. The peons, on the other hand, see themselves primarily as employees, not clients, and do not exhibit any particular loyalty or devotion to the landowners they work for. Likewise, godparent relations between peons and their employers are rare. Since access to resources for most households in Paso is still not dependent upon ties to wealthy landowners, the more traditional characteristics of patron-client situations have not yet evolved. Even if a landless family were cut off from access to their "patron's" land and employment, they would still have some other options open for survival and would probably be able to remain in Paso and find work. As the dependence of landless families on certain landed households becomes more crucial to their subsistence, however, it can be expected that peon-employer interactions will shift toward a more classic patron-client pattern.

While the economic power of the large landholders has been increasing in recent years, their political role in the community has been declining. This change reflects a shift in the basis of political power from within the community to outside it, and also a shift in community attitudes toward large landholders. Both in the past and in the present, Paso does not conform to Goldkind's findings in the central valley, which showed farmers who "hire labor on a permanent, year-round basis" to be the economic, social, and political leaders of the hamlet community (Goldkind 1961:374). Such economic power is definitely a factor in the high status of several medium and large agriculturalists in Paso, but some individuals who provide equal amounts of employment are of much lower status. Of the eight large landholders, only one can be said to hire labor year-round, but neither he nor any of the other seven can be said at this point to be recognized as leaders of the community.

The basis of power and leadership in Paso has been changing over the last 20 years from residing in the hands of those who control the community's productive resources (Fried 1960) to those who control relationships with powerful people outside the community (Wolf 1955). Such a shift is common in Green Revolution areas that have experienced changing relationships with outside institutions (Ruttan and Binswanger 1978). Twenty years ago, there were three powerful men in Paso, all of them large landholders. They were active in community service and derived power from these activities as well as from their wealth and control of land. These three leaders maintained ties with national political parties and were elected at different times as Paso's representatives to the cantonal council. At the same time, they were widely respected for their agricultural skills and hard work.

In recent years, the locus of power has shifted to alliances with national ministries and organizations. Of the men who could be called "powerful" or "influential" in Paso today, none is a large landholder. These new leaders are all small and medium landholders and, with one exception, are widely respected in the community for their mastery of the complex technology of tobacco production. These leaders are not listed by Pasanos who wish to point out "the rich" or those who live the most affluently; those would be large landholders. Each of the primary government agencies that operates in Paso deals primarily with one of these leaders. One man acts as a broker with the Agricultural Extension Service and is a member of the Cantonal Development Council, working with the bank and the Ministry of Agriculture employees on credit and extension programs. Another man is the head of committees that deal with the Caritas food program, the Social Help Institute, the Communal Welfare Committee, and the Society of Saint Vincent charity. He previously headed the liaison committee with the Mobile Health Team as well. Another leader is primarily in contact with the Ministry of Education and channels into the community benefits from that source, such as the repainting of the school. The fourth leader is the elected municipal representative to the Cantonal Council and has considerable influence with national political figures. He is also a leader of the soccer team in Paso.

Each of these men is in a position to ask for favors from the ministries and programs he deals with, and he can take credit for any improvements his organization brings into Paso. These favors or improvements can be community-wide benefits, such as the new highway, the electricity line, and the new community center. They can also take the form of family-level aid such as a new house, CARE food or clothing, or free access to a doctor. Projects such as a coffee producers' cooperative or group purchases of fertilizer also become available to the community through the efforts of these leaders.

While Pasanos continue to regard large landholders with respect for their wealth, the small and medium farmers who excel in tobacco and who furthermore engage in community service are accorded higher prestige and influence. It is important to note, however, that these four leaders all are landholders and all have sons who have finished school. Thus, they have the resources to allow them time away from agriculture to attend meetings and carry out activities required by their liaison positions. Access to land and labor seems to be a prerequisite to political power in Paso, but the control of vast resources is no longer necessary.

The decline in prestige of the large landholders is linked with the increasing rigidity of stratification and the decline of egalitarian interaction patterns. In particular, criticism is a powerful force of social control in the community. One person resigned a committee position because "there had been criticism" of his actions. In several other cases, criticism from "the community" was cited as the reason for a certain action being taken or avoided. One woman who had worked selflessly for two days on a school *turno* was reluctant to take some of the leftovers from the food concessions because "perhaps they will criticize me." Only when assured by several people that she deserved some recompense for her hard work did she assent.

As I have noted earlier, the squeeze of land scarcity has led to complaints from landless families that "there is nowhere to rent." Pasanos of many strata were vocal in their criticism of the large landholders' decisions to put land into pasture or hold it in fallow. The failure of large landholders to respond to these criticisms shows a hiatus in the power of such egalitarian sanctions. The situation increasingly suggests that large owners are above the control or influence of other community members, and this separation from the rest of Paso is clearly one aspect of their decline in local power and influence. No longer are the wealthier families seen as harder workers or more able entrepreneurs; even a few middle-sized farmers talked of the injustice of the current situation and of the decline in the large landholders' authority and legitimacy.

IMPLICATIONS OF THE CHANGING PATTERNS OF STRATIFICATION

Paso provides an example of the early stages of the emergence of fixed stratification based on increasing competition over productive resources (Carneiro 1961; Dumond 1965; Harner 1975; Van Vekken and Van Velzen 1972). Egalitarian patterns of marriage and the increasing farm size over the family cycle fit a time of greater fluidity in resource access. Many social interactions at community events or in visiting still demonstrate this democratic theme in Paso. At the same time, consumption standards are rising, and dress, diet, housing, and furnishings all show an increasing polarization between the rich and the poor. The relative lack of hierarchical patron-client relations may change as more and more families come to depend exclusively on such ties for their survival. The changing profile of leadership in the community reflects both of the forces for change elaborated in Chapter 1. The greater "development" of the community, via government agencies and services from specialists, has led to these resources becoming increasingly important in community life. At the same time, the population pressure and the changing land uses of the large landholders have eroded the legitimacy of their private property rights, resulting in a decline in their authority in the community. The emerging new leadership, respected for their mastery of the labor-intensive technology of tobacco, reflects these two forces for change. Both the internal and external pressures point to greater stratification in Paso and an increasing polarization of the wealthy and the poor as the future unfolds.

These three chapters have presented the context within which Pasanos make their production decisions. As farmers seek to meet their families' needs, we can see that their choices not only determine the quality of their own lives but also feed back into the context itself, shaping it anew for their neighbors and the next generation.

4

Production Methods and Markets: The Four Crop Options

Edwards (1967) defines the process of decision making as having two preliminary parts: first, the options available must be clarified; then, the options must be ranked in some fashion. The purpose of this chapter is to present the four main agricultural options open to Pasanos—these are the ways that they can choose to make their livings. The succeeding chapters will explore how these options are ranked and the results of those choices. Readers familiar with agricultural methods in Costa Rica may wish to skip to Chapter 5.

As the data in Chapter 1 reveal, farming activities in Paso are dedicated primarily to four land uses: tobacco, grains (corn and beans), coffee, and pasture. Almost all households choose more than one crop; even those dedicated to corn production usually plant beans as well. The majority of farmers have a mix involving two or more land uses. The characteristics of these crops—their production methods, marketing institutions, and general resource requirements—provide the basis on which Pasanos can judge which options will best meet their household needs.

CORN AND BEANS

The predominant land use in the Puriscal area until very recently has been foodgrains—corn, beans, and rice. This land use continues to be important in Paso; over 100 of the 916 manzanas in the community are in foodgrain production, the largest single land use after pasture (see Table 1.4). Although the amount of land in grains has been declining in the last 20 years, the number of individuals planting it has not declined. Virtually all households engaged in agriculture plant corn, and many plant beans as well. Rice has declined in importance in recent years because the soils are losing fertility. Chemical fertilization for rice is not as effective as with corn, and rusts and other rice diseases have hurt production. Corn and bean production comprised 35% of all agricultural and livestock income in Paso in 1972, the largest share for any of the land uses.

The traditional corn-and-beans rotation in Paso has changed little over the past century. When planted in rotation with tobacco, however, corn and beans production methods are somewhat different. These variations will be outlined later in this chapter as part of the description of tobacco production.

Traditional corn planting begins with the preparation of the soil in the dry season. Agriculturalists who still have land in fallow and who rotate their grain lands each year begin by cutting the brush or forest with axes and bush-knives (*cuchillos*). The brush is left to dry and then burned. The trash is usually cleared off the field and the soil loosened with a machete (a broader-bladed, curved *cuchillo*) before planting. Less affluent Pasanos who must use the same corn fields every year also prepare the soil by loosening it with machetes. Then, when the first rains come, farmers plant the fields with a metal-tipped dibble stick, walking in contour rows across the sides of the hills, making a hole, dropping in usually two grains per hole, then stepping on the hole to cover it while making the next hole.

Most Pasanos fertilize their corn. Financial considerations—cash on hand, availability of credit, the family's financial situation, and the price of fertilizer—determine the quantity of fertilizer used. Those who fertilize their corn most carefully will do so right after planting and then again when the plant has reached about four feet, "to give strength to the ear." A different formula of fertilizer is used for the two applications. Farmers who cannot afford to fertilize twice apply only the second fertilization. In Paso, fertilizer is applied by hand, pouring a lump of fertilizer in a hole near the sprouting plant. The hole is made with the dibble stick, about six inches from the corn sprout, and is sometimes covered over afterward with dirt. If the fertilizer were scattered loosely or mixed in with the soil, the heavy rains in this region would immediately leach it out. Also, if the fertilizer were applied near the plant, but little or no rain were to fall immediately after, the corn would be burned.

After the first fertilization, when the corn has reached six to twelve inches in height, Pasanos weed carefully. Machetes are used, chopping under the roots of the weeds, turning the soil over. As with most tropical soils, the layer of topsoil is shallow, and machetes are more suitable for cultivation in such an area than hoes, which would bring up the subsoil below. Farmers with steep plots work from top to bottom of the hill, but flatter land requires stooping to work it. A second weeding takes place when the corn plants are about a month old and roughly four feet high. The weed-free ground is then ready for the second fertilization.

When the corn is ripe enough to eat as corn-on-the-cob (*hilotes*), Pasano families harvest scattered plants. Though *hilotes* are taken randomly throughout the field, there is no evidence that farmers are consciously thinning clusters of corn. *Hilotes* are eaten boiled off the cob or are shelled and ground in various forms; the taste, different from regular corn, is widely enjoyed. Especially for the poorer families who may have run out of corn from the previous harvest, *hilotes* are a much-awaited delicacy.

When the ears are fully formed and ripe, Pasanos then double their corn to dry in the field. The stalk is partially cut around shoulder height and the upper portion of the stalk pulled downward so that the ear is pointing to the ground. The vulnerable tip of the shuck, where the pollen tassles emerge, is then pointing downward, and the rains are shed easily from the base of the shuck. The corn is usually doubled about midway in the rainy season. Most years, there is a brief dry spell at this time, the *veranillo* ("little summer"), which is a convenient time to double corn and plant beans. When the rains resume, the corn dries in the sunny mornings and remains on the stalk until the dry season begins.

Beans are planted by the time the corn is fully grown and doubled, when a low weed growth has covered the field. After doubling his corn, the farmer broadcasts beans into the weeds, then chops the weeds down with a machete or *cuchillo*, covering over the seeds. The weeds then form a kind of mulch and fertilizer as they decay over the sprouting beans. The beans may twine up the corn stalks, but most varieties spread out between the corn plants, growing only a foot or so in height.

Beans are not weeded; therefore the farmer's next task is harvesting at the optimal time. Ripe beans are quickly lost if they get too wet, because the pods crack open and the beans sprout or rot. With luck, a farmer's beans will be ripe just as the rainy season ends, so that they can dry partially in the field. In harvesting, bean plants are pulled up by the roots and left to dry in the sun. Piling the dried plants on a cloth, the farmer then threshes them with two poles. The threshed plants, separated from the beans, are usually dumped as garbage at the edge of the field, though at least one farmer spreads them back on his field, recognizing their value as fertilizer.

To prepare beans for storage or sale, they must be sorted by color and dried in the sun. This work is often done by women, though the whole family may help. Until this task, all the agricultural work involved with grains is usually done by men. Corn is harvested after the beans and is often brought to the house in sacks of ears still in their husks. The ears may be stored in this form in granaries near the house or shucked and shelled for sale. If the corn is still wet, it is dried in the sun near the house. Both corn and beans are transported and sold in burlap sacks.

Agricultural extension workers have introduced several changes in corn production. Ten or eleven years ago, a new variety of corn was introduced which was in all ways substitutable for the traditional varieties but produced more per manzana. Its use spread slowly throughout the community as neighbors shared its seed with each other. This variety, called Rocamex, has become interbred now with local varieties, and Pasanos who use "Roca" quickly point out "but now it is hardly Roca anymore." Only one farmer has recognized that buying new seed from town can remedy this deterioration. Even otherwise sophisticated farmers say that once you find a kind of corn seed you like, you keep it for the rest of your life.

New methods of planting corn have been less readily adopted. Agricultural experts advocate closer distances between corn plants and between rows and also recommend a fertilizer application underneath the corn seed as it is planted. Many farmers feel this process might improve yields but have not changed their planting habits. They react as a traditional peasant might be expected to: "This is the way we have always planted." But they do point out some of their misgivings about the changes. Closer planting distances will make it difficult to weed and double the plants, they say; the work space between rows will be too cramped. The close distance between plants may encourage pests and diseases to spread more easily. Closer planting requires more seed and takes more nutrition out of the soil, which then requires more fertilizer. Farmers feel that the soil cannot support that much more corn; part of the corn would grow poorly.

As for adding fertilizer at planting time, one Pasano pointed out that the new procedure would require five different operations to be carried out on each row—making the hole, adding a dab of fertilizer, covering it with a bit of earth, dropping the seeds in, and then covering the hole. Each operation would require its own trip down the row and the extra four trips were seen as too much more work for the benefit gained over the traditional one-trip method.

The sale of grains is carried out in an atomistic local and national market. Some Pasanos sell grains to others in the community who have run out or do not produce their own. These buyer-seller ties are not exclusive; they are similar to the flexible market for agricultural labor noted in Chapter 2. One widow who lives alone and whose land is entirely planted to coffee buys corn from many neighboring households on different occasions, maintaining diverse ties and depending on no one house exclusively. This pattern was also observed for other households who occasionally buy corn or beans.

Grain sales in the past were made primarily to local storekeepers in Paso or to travelling *comerciantes* who would come with ox-carts to buy corn, beans, and rice. Former grain wholesalers say that their trade used to provide excellent profits. With the advent of the all-weather road, however, farmers selling a small number of sacks can now load them onto the regular passenger bus and take them to the Puriscal market themselves. Sometimes the bus owners arrange for a truck to carry cargo on a market day, or the farmer can arrange with friends to pay a truck to haul their produce jointly. *Comerciantes* also come independently by truck into the peasant community the evening before the market, and many Pasanos feel their prices are high enough to warrant sale to them. If dissatisfied, the agriculturalist can always ship his grain to market and see how prices are there. Once at market, however, his grain must be sold; no one ever brings his grain back home again. In the town market, atomistic relations also predominate; there are no dyadic ties between buyers and sellers, such as noted in Haiti (Mintz 1967). Pasano grain sellers seek out the prices offered by competing grain buyers (individuals and companies) and sell quickly or leisurely, as they choose. Pasanos do not discuss marketing as a problematic aspect of grain production. They are generally aware of current prices, but no swindle or particularly bad luck with prices was reported during my year of fieldwork. In contrast, marketing problems with other crops were avidly discussed.

Prices for corn and beans during the last 20 years are shown in Figure 4.1. The price of beans has risen much more steeply in recent years than the price of corn. Corn has increased in value roughly 50%—from the lowest price of ₡22 per hundredweight to the high in 1972–73 of ₡32 per quintal. In contrast, beans have increased in value by 75%, though this figure glosses over the dive in bean prices in the mid-sixties. The prices in Figure 4.1 are the minimum prices set by the Costa Rican government, in order to maintain a floor for grain prices. Since the government will always buy grains from farmers at these prices, private grain wholesalers buy at a price slightly higher than this national minimum. While the government price of corn was ₡32 per hundred pounds in 1972, private buyers offered ₡34 and up to ₡37 at the Puriscal market. Hence, actual prices available to Pasano farmers are somewhat higher than the graph shows, but the annual fluctuations match, since the market price follows the government prices closely. This program has been in existence and functioned smoothly for a long enough time to remove from grain prices much of the uncertainty that farmers experience with other crop prices.

In assessing these price changes, it is important to evaluate the overall fluctuations in consumer prices. The urban cost-of-living index during the same 20-year period reflects a 50% increase in consumer prices (Dirección General de Estadística y Censos 1972). This indicates that from 1952 to 1972 there has been roughly 50% general inflation in prices in Costa Rica's urban areas. Unfortunately, price indicators for rural areas such as Puriscal are not available. There are reasons to suspect that inflation may be lower in rural

FIGURE 4.1. GOVERNMENT-SUPPORTED FLOOR PRICES FOR CORN AND
BEANS: 1952-1973 (IN COLONES PER 100 LBS.)

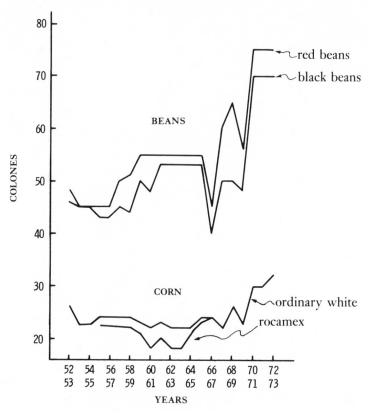

SOURCE: Consejo Nacional de Producción, San José, Costa Rica

areas, but the exact figure is not necessary to support the conclusion that the
prices of grains in Costa Rica have risen little if at all in real terms. These
price fluctuations, of course, have more effect on Pasanos who are producing
corn and beans primarily for sale. In fact, however, 52% of all corn produced
in the year of this research was sold, and 27% of all beans. Rice production
and sales were negligible. These figures indicate that foodgrains are produced to a considerable degree for family consumption and should not be so
strongly affected by price changes or cash crop production.

COFFEE

The amount of land in Paso dedicated to coffee is not large—71 manzanas out
of 916. In contrast to grains, however, this total has more than doubled in the

past 20 years. While occupying only 8% of the land area, coffee contributed 13% to the total income from agricultural production in 1972. Coffee is definitely a minor crop for most farmers, but 47 out of the 77 households maintain a stand of coffee trees. Pasanos enjoy coffee production and say they would like to dedicate themselves to it more than they do.

Coffee is the only permanent tree crop (other than fruit trees) grown in Paso. Coffee trees are started by planting the sprouted coffee berries in specially prepared seedbeds. The small plants are then transplanted to the fields where they will develop. In the first year of their development, they are sometimes intercropped with corn or tobacco—one man even planted cabbage between the rows of coffee. Intercropping is an attempt to get some return from the land while the coffee grows, but some crops rob the coffee of needed nutrients and slow its growth. Second-year coffee in Paso often produces well enough to repay the farmer for his cash costs for the year. By the third year, the coffee is producing well. In the past, coffee stands were shaded by *guineo cuadrado*, a relative of the banana tree, whose fruit was used to fatten the family pig. With the advent of the disease *moko*, most Pasano fields are now left unshaded.

Once the plant is established, coffee requires relatively little labor to maintain it. Traditionally, coffee producers in Paso did not use fertilizer, and there are still some older Pasanos who let the soil's own fertility suffice. The vast majority of farmers in Paso nowadays use chemical fertilizer, and all of the newer stands of coffee are carefully fertilized and pruned. Many Pasanos obtain this fertilizer on credit from the company to which they sell their harvest; their loan is subtracted from earnings with each sale of fruit.

Coffee trees flower with the first rains and then slowly develop their fruit—red berries with the coffee bean (seed) inside. In August and September, the green berries start to turn red and the coffee harvest begins. The fruit on each branch matures at different times, and only the ripe fruit may be sold. For this reason, each tree must be picked several different times. One large farmer states that his trees are picked over eight separate times. The coffee harvest, then, is a labor-intensive time. If the ripe berries are not picked, they fall to the ground and rot and are thereby lost.

The labor investment in coffee is high in the early years—considerable weeding, fertilizing, and pruning are necessary. Once the coffee is established, however, it requires perhaps three weeks of weeding, fertilizing, and pruning per manzana per year, and the work is considered "light" and "pleasant" by Pasanos. Then follows the coffee harvest, which requires considerable time but is also pleasant work. Whole families spend the morning picking coffee, and, if it does not rain, they continue on into the afternoon. Except for the harvest, all other agricultural work on coffee is usually done by the men.

The coffee harvest allows boys and girls to mix informally and also allows young men and women to earn some cash. Coffee harvest in the central valley is traditionally a happy social occasion (see Biesanz and Biesanz 1944). Single people are more likely to socialize on the Meseta Central than in Paso, where only 17 of the 47 households with coffee hire help from outside the family, but the chance to work outside the house is one which is welcomed by many young women. Each year two or three different Pasanos migrate to the Meseta Central to pick coffee and return in November or December. The migratory workers usually are unmarried sons, though one year an entire landless family went.

Coffee is the only crop in Paso that will produce some harvest with no attention, farmers say. One Pasano has a stand of old and poorly producing coffee. He has no grown sons and is very busy with tobacco and corn. While he does almost no work on his coffee, it still manages to produce a small amount without fertilizer or pruning. This farmer would like to replant and improve the coffee he has now and eventually expand his plantings to rely more on coffee production. For the present, however, he feels the high cost of hired labor, the price fluctuations of coffee, and the low return for the first two years leave him with no choice but to neglect his coffee. He says that the low risk and easier work of coffee make it attractive to him, and when his sons are grown, he may leave tobacco and move into coffee.

Most families who produce coffee save enough for their own consumption, and a few store it to pay peons in kind. To be stored throughout the year, coffee must be dried in the sun, so that the fruit will dry and separate itself from the seed and will not rot during storage. Some families pound the dried seeds with a wooden mallet and then winnow to separate the coffee beans from the dried fruit covering. These families then store only the beans. Coffee given in payment to peons is usually dry but unwinnowed.

To prepare coffee for family consumption, a woman of the house toasts the winnowed beans on an iron griddle and grinds them, producing ground coffee similar to that sold in stores. Both men and women are responsible for drying the coffee berries; men usually do the pounding and winnowing and women, the toasting and grinding. Some coffee producers prefer to sell all their coffee and buy bags of ground coffee at the store for their own consumption.

Ninety to ninety-five percent of the Paso coffee harvest is sold to one coffee company, which maintains a *recibidor*, or buying station, in the center of the community. This buying station was new in the year of my research and had been built by the coffee company after a petition was brought by a committee of Pasanos. Before there was a *recibidor* in Paso, coffee had to be transported to the *recibidor* in a neighboring community. Over the years, buying stations have moved progressively farther out from the town of Puriscal. Before the highway was built into Paso, coffee in small quantities was sold by carrying it out on horseback. In large quantities, ox-carts were needed, and some large farmers preferred to dry most of their harvest, then transport it dry. In either case, before the highway, very few households grew large amounts of coffee because of the difficulties involved in transporting it. Today, members of households near the buying station carry the sacks on their backs. Other families load them on horses, and a few families use the local bus to transport a sack or two each day.

Coffee, if sold wet (the entire fruit), is sold the day it is picked because it will otherwise quickly rot. Children most often bring the day's harvest to the buying station, usually in the afternoon. The *recibidor* is staffed each day by an employee of the coffee company who helps the farmer dump the coffee into a container where it is measured by volume. The amount of coffee received is recorded on a receipt, which farmers can redeem in cash at the point of sale, or later in town, according to the desired payment plan. At the end of each day, the coffee is loaded onto a truck and taken to the processing plant on the Meseta Central.

Many Pasanos measure their coffee at home first to keep track of the measurements made at the *recibidor*. The first employee who staffed the Paso buying station was said to undermeasure the coffee brought to him; repeated complaints at the coffee company office in Puriscal resulted in his replacement. As I left the field, the problem had recurred and Pasanos had begun to complain again. They were confident that complaints on different occasions by different households would result in a fairer employee. They did not feel the low measurements were made on purpose; lack of skill was felt to be the cause, since the employee's salary was independent of the amount of coffee he measured. These measurement discrepancies were the only problems Pasanos experienced with the sale of coffee.

Whereas the changes in marketing infrastructure in recent years have increased the convenience of coffee production in Paso, the price structure has fluctuated quite erratically. In Figure 4.2, the postwar period shows a steadily increasing demand for coffee and a tightening supply, as Brazil's coffee was badly damaged by frosts. These trends boosted prices sharply in the early fifties. All over Latin America, farmers planted coffee, and soon the increasing supply brought a "bust" in coffee prices. At this point, both the consumer and the producer nations agreed to regulate prices, and the first coffee cartel agreements were signed in 1962. There has been some relative stability in world coffee prices since then, subject to uncertainties every five years, as the agreements are renewed. In 1972, the consumer nations refused to agree to control prices, and when this research was terminated, coffee prices were uncontrolled and open to market influences. Figure 4.3 presents the quantities of coffee exports from Costa Rica over this same period.

Within Costa Rica, coffee export is carefully controlled. Each year, the governmental Office of Coffee sets quotas for all private export businesses, stipulating the percentage of coffee they purchase that may be sold abroad.

FIGURE 4.2. COFFEE EXPORT PRICES: 1942-1971 (IN DOLLARS PER 100 LBS.)

SOURCE: Oficina de Café, San José, Costa Rica

This mechanism assures a domestic supply of coffee. Companies compete to buy coffee from producers, and prices to the farmer may fluctuate slightly during the harvest season from this competition. When the harvest is complete and companies have arranged for their sales abroad, the Coffee Office reviews their books for adherence to the export quota and fair practices. Then the Office subtracts taxes, operating costs, and 9% profits from the gross income of each company. The remaining "profit" is then divided equally among all the producers who have sold coffee to the company, in proportion to the amount sold. Depending on world prices, this rebate (*liquidación*) may be minimal or as much as 50% of the original price paid to the producer. In this manner, the profit level of coffee companies is con-

FIGURE 4.3. EXPORTS OF COFFEE FROM COSTA RICA: 1952-1971
(IN WEIGHT PER 100 LBS.)

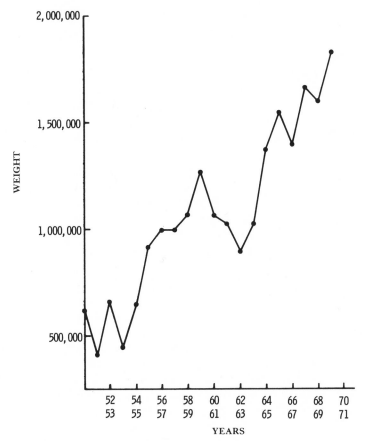

SOURCE: Oficina de Café, San José, Costa Rica, 1973

trolled by an independent branch of the government, and farmers participate
equally in the high profits accruing to high world coffee prices.

Although coffee prices have varied considerably in recent years, and are
currently lower than the 1950s high, Pasanos are increasing their coffee
plantings. Access to marketing infrastructure and the long-term price trends
clearly have more weight than short-term price fluctuation in determining
coffee plantings. Unlike the Paez coffee farmers of Colombia discussed by
Ortiz (1967), in Paso the decision to plant a stand of coffee is not made once or
at most twice in an agriculturalist's life. The Paez Indian reservation is a
situation of fixed land resources and almost the only way to obtain land for

coffee is through inheritance, which is usually received at marriage (Ortiz 1967:214). This situation provides the newly married household head with his lifetime land resources at the age of 20 and militates against reevaluations of land allocations to coffee.

In contrast, allocations of land to coffee are potentially evaluated at many different points in the life cycle of a Pasano household. The Costa Rican 20-year-old man is rarely married, and he will continue to aspire to increasing his landholdings until old age. With each new land acquisition or sale, the Pasano can reevaluate the proportion of his land in coffee and change allocations accordingly. New coffee is usually planted in small sections over several years to spread the labor cost. While no case was ever reported to me of good coffee being cut down in Paso, farmers did not hesitate to cut down old coffee that was not producing well and replace it with new coffee or change the land use to another crop. Thus, farmers weigh the characteristics of coffee—relatively low labor input, easy and agreeable work with sharp peaks of demand, necessity for cash expenditure for fertilizer, and relatively low risk—more frequently than do the Paez Indians, but perhaps less frequently than for the annual crops of corn, beans, and tobacco.

TOBACCO

The increase in tobacco acreage has been one of the two major changes in land use in Paso. Since tobacco is labor intensive, the average plot size is only 1 manzana, giving a total for Paso of 45 manzanas in tobacco in 1972. While the absolute acreage in tobacco is small, its growth rate is the highest of any in the community, and the importance of tobacco far outweighs its small acreage. Tobacco made up 23% of the total income generated by agriculture and cattle production in 1972, and when the value of the corn and beans that follow tobacco is included, this land use made up 34% of the total community income.

To grow tobacco, a Pasano must first obtain a contract with a tobacco company. The contract stipulates the amount of land the farmer intends to plant in tobacco and the estimated weight of his production in quintals (100 pounds). Farmers conceive of their contracts by weight: "I have a contract for 10 quintals," meaning 10 hundredweights, or 1,000 pounds of tobacco. In addition, the contract indicates the prices the company guarantees for each class of tobacco. A contract also assures the farmer that he can obtain credit, either from the tobacco company or from the bank. Since the cost of tobacco production is high, credit is a necessity for most households. Once a farmer has a tobacco contract, he is obligated to sell his tobacco only to the company he has contracted with, and conversely, the tobacco company is obligated to buy all the tobacco he produces, with the proviso that it be of acceptable quality.

Usually, in order to obtain a contract for the first time, a farmer must cultivate a small plot of tobacco on his own and sell the trial harvest to the tobacco company. If it is satisfactory, he will be advised to put his name on the list of applicants for the following season. Sons who have worked with their fathers on tobacco may be able to get their own contract based only on that association. Some companies, however, do not require any evidence of ability to handle tobacco but will give a contract to virtually anyone who applies. Once a farmer has a contract, he will be able to renew it each year automatically unless his crop was so inferior that the company decides to take his contract away.

Tobacco companies offer contracts without reference to the amount of tobacco to be planted; it is the farmer's decision to ask for "five quintals" or "ten quintals." The size of his contract determines the ceiling on his credit, however, so he would be unwise to ask for five when he needs money to produce ten. On the other hand, if a man has a ten-quintal contract and produces only eight quintals, he will be unlikely to get a contract for ten the following year; the company will probably require him to stipulate eight or fewer quintals. To be on the safe side, Pasanos prefer to slightly underestimate their capabilities. A man with a ten-quintal contract may cultivate enough plants for twelve or even fifteen quintals, given good weather. If it is a bad year, he will be sure of getting at least ten and not having to lower his contract. If a producer brings more tobacco to sell than his contract stipulates, the company will always buy it; only if he underproduces his contract will there be unfavorable consequences.

There were four tobacco companies in the Puriscal area in the year of my research. Two were long-established cigarette companies that give contracts for *criollo* (native variety) tobacco. Their operation is regulated by a semiautonomous branch of the Costa Rican government called the Tobacco Defense Board. In order to stabilize tobacco production and prices, the government controls these companies so that the amount of tobacco produced annually in Costa Rica does not exceed the annual consumption of tobacco in cigarettes. Competition between producers is reduced, and the companies do not maintain large stockpiles of tobacco. Contracts with these two companies are difficult for Pasanos to obtain, since the total quantity of tobacco each company can purchase does not increase much each year. The Puriscal area is one of three tobacco-producing regions in Costa Rica and has benefited in the last few years from an increase in tobacco diseases in one of the other tobacco regions. Therefore, during the period of this study, Puriscal's share of tobacco contracts has been expanding.

The other two companies in the area buy *cubano* (Cuban variety) tobacco, for cigars. They are export companies, shipping either tobacco or cigars to the United States. The United States market for foreign tobacco has been increasing in recent years: the older of these companies had been operating for seven years and the other began in the year of this research. The older *cubano* company is actually a cooperative of tobacco producers, which was set up under the guidance of Cubans. Unfortunately, the management of the coop has suffered from poor investments and fraud, and many of the Pasanos who have worked with the coop feel it is a failure. A new administration attempted to salvage the enterprise, but when this research ended the coop had not given contracts for the succeeding year and seemed to be defunct. The remaining *cubano* tobacco company has taken over most of the coop's clientele and is recruiting many new producers as well. Both *cubano* companies were anxious to expand and would generally give contracts to any interested farmer.

Credit is an essential part of the tobacco crop option. In the past, tobacco credit was always administered through the tobacco companies; they arranged for fertilizer and insecticide in appropriate quantities for each producer. When the contract holder presented his crop for sale, his debts were subtracted from the value of his harvest and he received the balance in a check. In the year before this research, a new policy of funding small farmers was implemented by the banking system in Costa Rica (which is nationalized and directed by the Central Bank): the tobacco companies have allowed the government to take over the costs of the administration of tobacco credit. Under the new procedure, a farmer receives a contract from a tobacco company and then uses that contract as collateral for a loan from the bank. The bank arranges for him to withdraw fertilizer and insecticide from local supply companies (just as he used to do with credit directly from the tobacco companies) and gives him cash if he needs it for hired labor or for home consumption. When the tobacco company receives the producer's tobacco, it pays him by a check that must be cashed at the bank, where bank officials deduct his outstanding loan and interest and pay him the balance. Tobacco credit in this form is very similar to the former arrangement from the Pasano point of view. It requires more time waiting in line at the bank, however, and for this reason a few Pasanos prefer to use their own credit at the fertilizer agencies and bypass the bank entirely. In any case, interest is 8% per year, and there have been no defaults on loans or credit for tobacco in Paso (see Chapter 8). While Pasanos regret the necessity for credit and dislike the state of being in debt, the benefits they gain from cultivating tobacco outweigh these drawbacks for the majority of households.

Tobacco in rotation with corn and beans is carefully adapted to the Puriscal climate. Tobacco has a short growing season and is planted toward the end of the rainy months. It matures about when the dry season begins, and its leaves dry on the warm, sunny days. Tobacco is harvested and the fields prepared before the rains begin again, when the farmer then plants corn and beans in alternate rows. The beans ripen first and are harvested, followed by the corn, which is ripe by the middle of the rainy season. The corn is harvested in time to prepare the fields for tobacco once again.

In order to plant tobacco, the fields must be terraced. The soil is ridged in contour lines, a process that carves the sides of the hills into steps. On flatter land, the terraces may be only a foot or so high, but on steep slopes, each ridge may be two or three feet below the one above. Terracing is done with a shovel that is sharpened to a knifelike edge. One farmer states that he moves across the hill at least four times to make each terrace, throwing shovelfuls of soil from a trough to a growing ridge. The labor is very difficult, and many *tabacaleros* hire help for the terracing. In the process of ridging, the earth is hilled up over the corn stalks and debris from the preceding harvest. In this way, there is some organic decomposition, which improves the fertility of the soil.

When the Pasano farmer begins his tobacco terracing, he also cultivates a small plot for a seedbed. Tobacco seeds are fertilized and watered occasionally and may be sprayed with insecticide as well. When a section of terrace has been completed, the family transplants the tobacco plants to the terraces, by hand. The sight of Pasanos pulling up the six-inch-high seedlings and stooping to plant them along the terraces is very reminiscent of paddy rice transplantings.

Tobacco production involves a higher level of technical expertise than any other crop. The tobacco is fertilized a day or two after planting. The same procedure is used as with corn and coffee—a hole is made in the ground near the seedling and fertilizer poured into it. In the three- to four-month growing season, tobacco is fertilized three times, all with the same formula. This massive fertilization accounts for most of the high cost of tobacco production. In addition, tobacco must be sprayed for insects, worms, and diseases. Insecticide is sprayed with a pressure pump worn on the farmer's back. The sprayers represent another significant expense of tobacco. It is possible to borrow a sprayer, but a delay in obtaining it may result in serious leaf damage, lowering the quality of the crop. For this reason, most Pasanos find it necessary to own their own sprayers. During the growing season the tobacco plants must be pruned in several different ways to form the bush to a shape for maximum weight in the leaves. Each plant must be pruned several times, and although the work is easy, it is time consuming. As with all the tasks of tobacco production, considerable skill is necessary. Timing and method of fertilizer application, insecticide, and pruning all require experience and methods that are slightly different for each plot's slope, soil, tobacco variety, and so on—the farmer must be sensitive to the microenvironment of his fields. Experimentation to develop the best method for each plot takes time, and more experienced tobacco producers, as will be shown, have higher profits.

When the leaves are fully grown and ripe, the *tabacalero* picks them one by one, by hand. As with coffee, each plant must be harvested several different times. The leaves are then tied two at a time to drying poles. Cuban tobacco is placed in a drying shed to dry in the warm air. *Criollo* tobacco is dried in the open sun, then placed in the drying shed. A drying shed usually has a tin roof with a framework of boards inside to support floor-to-ceiling rows of tobacco. Many households thatch the outside with palm leaves to protect the tobacco against the wind or a surprise rain. This drying shed is another costly investment required by tobacco.

Throughout the growing season, the tobacco producer is visited by inspectors from his tobacco company and from the Tobacco Defense Board. Inspections usually take place every two weeks, and the company officials advise farmers about problems they are encountering or about the correct timing and procedure of their work. This advice is very welcome to less experienced workers and helps them avoid errors and losses. For the older producers, the supervision is more likely to be annoying. Experts do not always recognize differences in microenvironments, and sometimes Pasanos feel they know better than the experts. While the farmers rarely disagree openly with the supervisors, they may ignore their instructions. One farmer, for example, felt he had been advised to harvest his tobacco too early before the leaves had reached their maximum weight. He delayed the harvest for two weeks and began to pick leaves just before the inspector's next visit.

When the harvest is complete and dry, the leaves are untied from the poles for classification. Both the string and the poles are saved for the following year. Tobacco is classified by several factors, and some companies require as many as nine different categories. As the tobacco is sorted, bundles of the same class are tied together by their stems. In this form, the tobacco is carried to Puriscal on an assigned day and sold to the tobacco company.

The sale of tobacco is one of the most problematic aspects of tobacco production. The farmer has no control over how his tobacco will be classified when it is presented to the tobacco company. Each bundle of leaves is inspected and thrown into a pile according to the judgment of the inspectors. A bundle that the farmer feels is first class may be thrown into second class, and a second-class bundle into third. The downgrading of tobacco is a frequent complaint of Pasano producers, though some companies are said to be worse than others. As a result of a bad experience with downgrading, one agriculturalist refused to continue working with his company and waited two years on the lists of another company before getting a contract.

Among its other purposes, the Tobacco Defense Board was set up to mediate in such situations. Many of the officials of the board were previously employed by a tobacco company, however, and the board members are also of a social class similar to the tobacco inspectors, a group clearly distinct from the peasant producers. The board seems to function effectively in protecting farmers from unwarranted termination of their contracts and in supervision of their crops, but protection from biased classification is a knottier problem. Sorting tobacco bundles fairly is a difficult task in any case, but the conflicting economic interests of the producers and the buyers make this situation very difficult. By law, farmers can appeal for arbitration to the tobacco board when they feel their crop has been downgraded. The few Pasanos who know of this law say they would never do so, however, because it would anger the company classifiers. Then, they say, the sale would go worse for them than it had before, and it might even jeopardize their contract.

Tobacco prices are another area in which the Tobacco Defense Board is expected to mediate the interests of the producers and the companies. It has been responsible for a number of increases over recent years, though farmers complain that prices are still far too low. Figure 4.4 shows that since 1962, when the board was established, tobacco prices have risen by 35% for first class. Given that urban consumer prices rose by 20% from 1964 to 1972, tobacco prices can be seen to be exceeding urban inflation by a small margin. Figure 4.5 shows that exports of tobacco have fluctuated since 1962; there were no exports of tobacco from 1952 to 1962.

FIGURE 4.4 DOMESTIC TOBACCO PRICES: 1962-1973
(*CRIOLLO* TOBACCO, IN COLONES PER POUND)

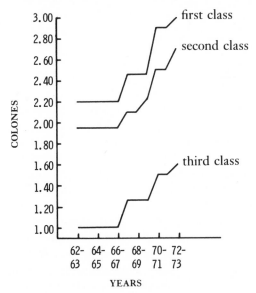

SOURCE: Junta de Defensa del Tabaco, San José, Costa Rica

FIGURE 4.5. EXPORTS OF TOBACCO FROM COSTA RICA: 1962-1972
(IN WEIGHT PER 1,000 LBS.)

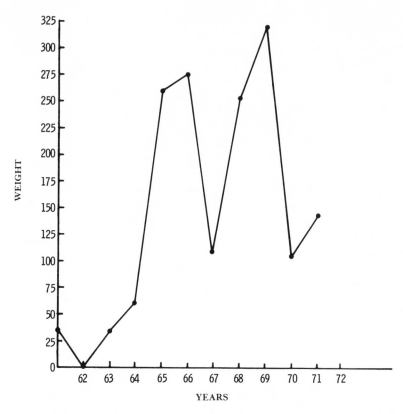

SOURCE: U.S. Dept. of Agriculture, Washington, D.C.

After the tobacco harvest, the farmer pulls up the denuded stalks and lays them lengthwise in the terraces. He then plants corn and beans when the rains begin. Usually the corn and beans are both planted with a dibble stick, in parallel rows (two grains of corn to a hole, and three to four beans per hole). Many Pasanos fertilize the corn, and incidentally the beans, once or twice in their growing season. The ripe beans are harvested and threshed before the corn is ripe. In contrast to traditional grain production practices, corn is not doubled and left to dry in the fields but instead is harvested ripe but wet. The grain is either left on the cob or shelled and dried in the sunny mornings of the rainy season. One farmer was annoyed with the increased bother of sunning corn and then covering it when it looked like rain. As an experiment, he planned to cut the corn stalks off and lay them over bamboo poles with the ear pointed downward, simulating the doubled corn in the field and, he hoped, yielding the same field-dried ears of corn.

When the corn and beans have been harvested, the agriculturalist uproots the corn stalks and lays them lengthwise in the terraces, as he did with the tobacco plants. Then he proceeds to build up the soil ridges again, moving the soil from the preceding year's ridges and hilling it up over stalks, creating a different line of ridges. In this way, the soil of the hillside is moved back and forth each year and is both chemically and organically fertilized. New terraces are hardest to work, but after several years of tobacco production, the soil is loosened and rocks are removed. The effort of carving up the hillside into dry terraces never becomes easy, but most farmers say the backbreaking labor increases the soil's fertility, and Pasanos value highly this characteristic of tobacco.

The division of labor among men and women for the corn and beans part of the tobacco rotation is the same as for traditional grains. In the production of tobacco, however, women and children play a crucial role. Except for terracing and weeding and spraying insecticide, the whole family can be seen working together on their tobacco. Women and children help to plant, prune, fertilize, harvest, tie, untie, classify, and retie tobacco. Since most tobacco work is done in the dry season, or summer, Costa Rican children do not miss school while helping in the fields. One man stated that he liked tobacco as a crop "because it lets me utilize the labor of my family." While women and children help with coffee harvests, too, their labor on tobacco represents longer hours and harder work.

The risk involved in the production of tobacco is greater than for any of the other land use options in Paso. In one harvest year, Pasanos suffered a wide array of problems with their tobacco: one man's seeds did not sprout and he had to reseed; another burned his seedbed with fertilizer. Caterpillars and worms ate holes in many farmers' leaves, while incorrect insecticide application killed both the insects and the tobacco in one case. The rainy season ended abruptly that year, hurting several families who had planted late. A sudden rain in the dry season knocked over more than one drying shed, wetting the stored tobacco and damaging much of it with water spots. It was a bad year for summer winds, and there was severe damage in a few cases where high winds burned the tobacco leaves off right down to the vein. One man's soil got a tobacco disease which made the plant "faint" and die. He made so little on his harvest that he had to hold his debts over until the following year. This year was not considered atypical by Pasanos; their discussions of the past suggest that if it is not one thing, it's another.

To balance this risk is the grain production in the terraces. Corn and beans produce extremely well in terraces and are not subject to as many fluctuations in productivity. All tobacco farmers discuss this characteristic of tobacco—"it gives me good grains afterwards"—as an important aspect of their decision to produce tobacco. In many cases, farmers barely break even from the sale of their tobacco, but the crop is worth their while from the good harvests of corn and beans that follow. (These comparative figures will be presented in Chapter 5.)

CATTLE

The other major change in Paso is the shift of land out of agriculture and into pasture. Twenty years ago, there were over 200 manzanas in pasture, but today there are 535, with more land added every year. Cattle and milk sales made up 29% of all the income generated by Paso's agricultural production in 1972. This shift has already been shown to have serious consequences for the economic options of the landless groups, who are now more pinched for land to rent. Cattle raising is almost exclusively an enterprise chosen by middle and large farmers, the strata who own land beyond that necessary to maintain an adequate standard of living from crop agriculture.

There have been a few technological changes over the history of cattle production in Paso. Considerable research by the Costa Rican Ministry of Agriculture and international agencies has been invested in grass types and cattle breeds. Older grasslands were planted with *pasto dulce*, a low, sweet grass native to the region. In recent years, *ganaderos* (cattle producers) all over Costa Rica have tried African grasses and found them superior for fattening cattle. In Paso, the most popular improved pasture grass is *jaragua*, an African import that grows up to three feet tall and fattens cattle better than *pasto dulce*. *Jaragua* is a tough, aggressive grass that quickly takes over a field if it is seeded into burned ground. *Jaragua* roots do not hold the soil, however, and on the steep slopes the cattle's weight contributes to serious sheet erosion. In areas long in pasture, the soil sometimes becomes packed very hard and gradually loses fertility. Softer slopes, of course, fare better and are more able to utilize the manure for improving fertility. A few Pasanos have also begun to plant *estrella africana*, a newer African grass which must be planted from cuttings. This labor investment deters most farmers, but *estrella* is noted for its elaborate root system, which "weaves" the soil and is less likely to cause erosion. The native Costa Rican cattle (brought by the Spanish colonists) have been crossed with Brahma strains from India, and many Puriscal cattle show these Brahma characteristics. Recently, French and other breeds are also of interest to beef producers.

Pasanos who feel the pinch of the land shortage fear that once the land is put into pasture it cannot be returned to agriculture. Experts report that it is possible but difficult. Pasanos who have tried it say that with *jaragua*, especially, the seeds continue to come up, and even ploughing will not solve the problem. *Estrella africana*, with its mesh of roots, is even more difficult to remove. Large landholders do not seem to be concerned with this problem, since most of them plan to keep their land in pasture. It might be possible, however, to reclaim the land by letting the pasture revert to forest and then cutting down forest for agriculture.

Cattle as a land use option is low in labor and capital costs. To maintain a pasture once it has been planted, Pasanos chop the saplings and competitive growth with a machete. One large landholder estimates that it takes two jornals, or twelve hours, to keep a manzana clear each year, but most households spend half that much time or less. Cattle are usually checked once a week for cuts and bites, and are fed salt. Some cattle are vaccinated once or twice a year, a minor expense in both time and money. Pasanos use barbed wire fences that occasionally must be replaced or mended. Thus, cash expenses involved with cattle maintenance are only salt, medicines, peons to weed pasture, and fences. The major cost, therefore, is the cattle themselves, and Pasanos who have large pastures usually build up their herds slowly by keeping their calves rather than selling them.

Cattle are sold both locally and in Puriscal. Some farmers drive their cattle or calves to market on horseback; others rent a truck, often sharing the cost with friends or relatives. During my year of research there was only one case of cattle sale between two Pasanos, but in the past, the large landowners more frequently acted as middlemen, buying up a number of cattle and then herding them all to market at once. Pasanos discussed no problems with the sale of cattle. Again, as with corn and beans, the seller has no ties with specific buyers in Puriscal but seeks the best price among a number of competing *comerciantes* in the market.

Families value cows for milk, and if their pasture is close enough to their house, they keep a cow and calf for this purpose. The purpose of cows is not primarily for milk, but rather for the production of calves, however much the family may enjoy the milk cows give.

The steep rise in pasture acreage in Paso correlates closely with recent prices in Figure 4.6, which show that the value of an average beef animal for export has more than doubled in the last 20 years, nearly doubling in the last 10 years alone. Compared to the cost of living, beef is the only land use option open to Pasanos that has greatly increased in value in spite of inflation. The price increase does not reflect world-wide inflation in beef prices, however. The wholesale price of beef in the United States was $30.26 in 1950 and $35.71 in 1972—an increase of only 18% in 22 years (U.S. Department of Commerce 1973:352). Since nearly all the beef exported from Costa Rica is for the United States market, the price increase in Costa Rica represents a real jump in the value of Costa Rican beef. This price increase is more than three times the increase in prices of any other crop discussed here.

Figure 4.7 shows the total Costa Rican export of beef for the 20-year period 1953–1972. Cattle exports in 1953–54 numbered 596 head; in

FIGURE 4.6. CATTLE PRICES FOR EXPORT AND DOMESTIC CONSUMPTION: 1952-1975 (IN COLONES PER ANIMAL)

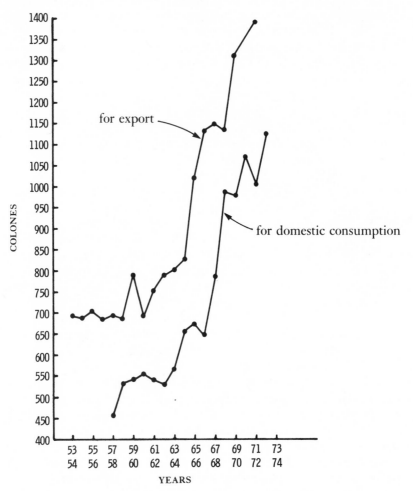

SOURCE: Consejo Nacional de Producción, San José, Costa Rica

1971–72 they totaled 110,303, all of which were butchered in Costa Rica. The national development of the beef industry parallels the growing interest in that land use in Paso, as well.

˘ SUMMARY

This review of the four land use options sets the stage for the analysis of the value of each of these options for different kinds of farmers in Paso. Burling

FIGURE 4.7. EXPORT OF BEEF CATTLE FROM COSTA RICA: 1953-1972

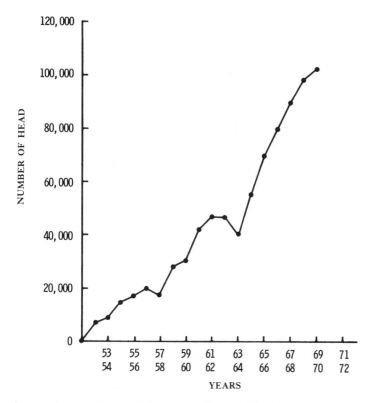

Source: Consejo Nacional de Producción, San José, Costa Rica

points out that, in making choices, options can be seen as both divisible and indivisible (1962:810). In this regard, Pasano land use choices are completely divisible—a whole manzana can be planted, for example, to pasture or to a mix of grains and coffee. The holdings of one household are usually divided among a number of options, and it is possible to choose all four options at once. Ortiz's work shows that some options are linked and must be chosen together (1967; 1973). In this Costa Rican case, however, all the land uses can be chosen separately, and none requires or eliminates another. Since almost all Pasanos have a mix of land uses, they clearly see their land as divisible and evaluate each option separately. This characteristic of Pasano land use allows the researcher to analyze land use decisions by asking: "Why does this farmer choose this crop for part of his mix?" Thus, the pros and cons of each crop option can be analyzed independently of other choices, exploring the utility of that option, as I have done in the next chapter.

5

Farmers' Cost-Benefit Analysis and the Tobacco Decision

Tobacco is one of the two major land use changes in Paso, and certainly the most complex of the new agricultural options. The purpose of this chapter is to explore the reasons why so many Pasanos have begun to plant tobacco, a crop virtually unnoticed 20 years ago. To understand this decision to pursue tobacco production, we will first look at Pasanos' statements explaining why the crop has become so popular. Their analysis that population pressure is causal will be tested with data from the community. Turning then to the actual decision process whereby farmers rank the four crop options, we will again compare quantified measurements of the important decision criteria (profit, labor and capital requirements, yields, and risk) with farmers' statements. These comparisons will clarify the utility of tobacco in a situation of land scarcity.

WHY TOBACCO?

Pasanos give a variety of reasons for the increasing popularity of tobacco. When asked why people are shifting land out of traditional grains and into this new land use, farmers say:

> Lots of people plant tobacco now because land is scarce and tobacco requires little land.

Tobacco is best on a small plot of land—one can get more from the soil because of the fertilizer and the terrace.

Tobacco pays on a little bit of land, and the corn after it is easy work.

I had to start tobacco because my [traditional] corn and beans were not producing at all.

Pasanos thus stress the high returns to scarce land and the yields of corn and beans after tobacco as the main reasons for the increasing acreage in this crop. This assessment implies that the growing population density and increasing land shortage have motivated a reevaluation of traditional criteria of land use decisions.

Such a process of population pressure and changing land use is noted by other researchers. "Only extreme population pressure, in terms of the capacity of each system, will force farmers to shift from swidden to more intensive patterns of land use" (Sanders and Price 1968:96). ". . . Practices of extensive agriculture are normally adhered to until population pressure becomes such that the system ceases to be viable through lack of sufficient land for rotation" (Dumond cited in Netting 1974:37). It has already been pointed out that few Pasanos have sufficient land to rotate with fallow and that most farmers must use chemical fertilizers to obtain even a minimal yield on their fields. Do the figures for population density, however, uphold their generalizations that land scarcity is the cause of intensification?

The population density of Paso can be measured in two ways. The first method uses the total population censused in 1972 together with measurements of land area taken from aerial photographs and the land use survey discussed in Chapter 1. By these calculations, Paso's population density is *201 persons per square mile*. This figure can be compared with the 64.0 persons per square mile of the Tsembaga Maring swidden agriculturalists of New Guinea (Rappaport 1968:33) and the 25.9 persons per square mile of the Hanunoo of the Philippines (Conklin 1957:10). Paso's 201 far surpasses these figures and approaches the density of the Kofyar of Nigeria, whose agricultural methods support 290 persons per squre mile (Netting 1968:110). The Kofyar, however, intensively manure, terrace, cultivate, and drain their fields, and practice crop rotation and water control as well.

A second calculation of population density in Paso reveals the urgency of the situation facing the majority of farmers. If the lands belonging to the large landholders and the lands belonging to the few absentee landholders are omitted from the computation, and the families of these large landholders are omitted as well from the total population, *each square mile supports 481 persons*! This figure is even more remarkable when Paso's rugged, mountainous terrain is taken into account. This second calculation leaves out the small plots of land owned by the large farmers and rented to the landless. To balance this bias, however, it leaves in all the land owned by the wealthier medium landholders, some of which is in pasture. Thus, the concentration of land resources in the hands of a few families creates a situation of land shortage and population pressure for the remainder of the community, and Pasanos' assessment that the increase in tobacco reflects a situation of land scarcity is correct.

We can now turn to the issue of Pasanos' evaluations of these crop options and the measurement of these variables. Readers uninterested in the technical details of emics and etics and the cost-benefit methodology can skip to p. 97. With an assessment of farmers' criteria for choosing tobacco, its value over the other three crop options in a situation of land scarcity will become clear.

The Evaluation of Crop Options—Emics and Etics

In discussions about agricultural choices, Pasanos' statements reveal that they use five main criteria to assess the crop options available: profitability, labor requirements, capital requirements, risk, and yields of grains:

> Tobacco gives you a lot of money, but then it costs a lot too and is risky.

> Corn pays, if you use fertilizer, though there is risk. Beans alone don't pay; the risk is too high.

> I like tobacco because of the good corn and beans that come after it.

> Even though tobacco is a lot of work, it's better than coffee, because coffee prices aren't so good now.

These statements indicate the emic criteria of choice. My usage of the "emic-etic distinction" follows Harris (1968), who distinguishes emic data as information that reflects the cognitive orientation of the people involved and uses their own units and categories for description. Etic data are objectively verifiable by an outside observer and are measured in units decided upon by the observer (Harris 1968:580). In assessing the crop options in Paso, it is important to compare farmers' ideas of which crops are most profitable or require the most work with my own measurements of these variables. These etic measurements are not designed to see if farmers are "right" when they say, for instance, that tobacco is more profitable. If these emic and etic measures diverge, then farmers may mean something quite different in saying "more profitable" than was originally understood, and this difference needs to be explored. In addition, the goal of the etic measures is to allow for the possibility that there are patterns in these variables that farmers are unaware of. Both of these purposes of the etic data will be met in the discussion below. It is expected that in some cases, emic and etic measures are identical—that is, my calculation of a farmer's profit in tobacco may be carried out in exactly the same way he does it. In other cases, the careful quantification of inputs into the production process may be quite foreign to the actual decision process of the farmer. Etic data were obtained by careful questioning as well as by observation, from the standpoint that actors can report on data in categories and for purposes foreign to them.

METHODOLOGY

To obtain the figures presented in this and subsequent chapters, I conducted detailed interviews with all households. These interviews usually lasted several hours, and in a few cases they were continued on two or three different afternoons. Farmers were asked to reconstruct all the costs and benefits from all their land uses for the previous year. In addition to this more formal interviewing, I also discussed farmers' opinions about the different crops and probed for the reasons why they had changed their land allocations to different crops in recent years. Issues of farm management and their decision-making criteria thus emerged in these discussions.

For the annual crops of corn, beans, and tobacco (and the minor crops grown by a few households, such as rice and broomcorn), I asked farmers to recall first all their cash and labor inputs and then the value of the harvest. They usually remembered cash inputs without difficulty: "I put on ten bags of fertilizer at ₡35 per bag." Transportation costs were also straightforward. Inputs that were not purchased, such as seed, were valued at their purchase price, though these costs were minor.

To reconstruct labor inputs accurately, I asked farmers to break the production cycle into a sequence of tasks and to calculate the number of days of labor invested in each task, distinguishing between unpaid family labor and hired labor, paid either in cash or in kind. Though a farmer might have spent a few regular workdays (jornals of six hours) plus several half days, he was able to calculate without difficulty the equivalent in whole workdays. This facility in labor calculations stems from the fact that Pasanos often conceptualize and discuss a task according to the number of jornals it would take a hired peon to do the job. For example, many individuals used the terminology "I spend six peons preparing the soil." Some farmers remembered in great detail: "How long did the weeding take? Well, from the road to the big tree usually takes me one peon [one day/one jornal] and then from the tree to the creek is another day. On the other side of the creek is at least two days plus a piece at the bottom of the hill is another. That's five peons for weeding." Labor of women and children was evaluated by the household head and indicated as the number of adult male equivalents. This list of total labor and recurring cash costs was then summed in two parts, the total paid costs and the total unpaid costs.

Pasanos indicated that their own profit calculation does not attribute a wage to unpaid family labor: "Oh, no, don't figure in the time we spend working," they said. "We don't come out making anything then." Another farmer said, "I can't count my own work on tobacco or the children's either; you don't make money on tobacco if you count the family's work." These farmers are aware of their labor investments but are making decisions based on "the annual product minus outlays" (Chayanov 1966:5). I refer to this kind of cost-benefit analysis as a Chayanovian calculation.

The "annual return" or gross profits of crops like tobacco and coffee were easy to determine because they were sold. For crops not sold, such as corn and beans, the harvest was valued at the price it would have brought had it been sold at the time of harvest. Chibnik (1978) argues that such a procedure undervalues crops produced for household consumption, and he indicates that a more accurate figure would be the consumer's *buying* price. In Paso, however, families who run out of corn do not always buy it, and there is some evidence to suggest that if the corn harvest is good, corn consumption rises. Valuing the harvest at the consumer price therefore involves an assumption of constant eating habits that is not supported by the data. Most householders do sell part of their corn production, and therefore this price is used in the cost-benefit calculation.

An additional category of costs emerged in these discussions: the fixed costs of tobacco sprayer and shed, agricultural tools, and the investment in fences for cattle. While the planting of coffee bushes could also be seen as this same kind of capital investment, Pasanos never discussed these labor costs as an investment beyond noting that profits are low in the first few years of coffee because yields are low. The way these costs and the costs of cattle purchased for breeding or milk are valued depends on the kind of cost-benefit calculation undertaken. Elsewhere (Barlett 1980b:6), I have compared different cost-benefit methodologies and shown that traditional economic calculations, based on accounting principles developed for capitalist firms, founder on the issue of opportunity costs (the return from resources at their next best use). Especially when calculating the appropriate "cost" of unpaid family labor, researcher bias can seriously distort the usefulness of the measure. Since my goal is not to evaluate whether these farmers are behaving as if they were capitalist firms but rather to understand how they weigh the different crop options at hand, the cost-benefit methodology that distorts the least is most desirable.

In a sample of Pasano households, the traditional cost-benefit calculations were undertaken using the following methodology (see Barlett 1980b): all labor was valued at the going wage, tools and other fixed costs were depreciated over their expected lifetimes, and the harvest was valued at both its sale price and its consumer purchase price, according to the percentage of the crop sold. These calculations resulted, of course, in a much lower profit figure than the Chayanovian calculation, but, more important, these traditional cost-benefit calculations did not change any of the conclusions based on the Chayanovian calculations as to the relative utility of the crops. In one aspect, however, the traditional economic calculations did distort farmers' decisions. By attributing a fixed wage to family labor, the calculation suggests that some farmers actually lose money. In fact, these households have many workers and are investing more labor in the household enterprise, since that labor is available. Thus, the opportunity cost of their labor is too high, when set at the going wage. Since alternative opportunities for labor vary from family to family and from month to month, as well as according to the age and strength of the child and the contacts with neighbors who might hire him or her, in reality the exact figure most appropriate to understand each household's decisions is very hard to determine. Therefore, profit is best calculated by the Chayanovian "returns to labor" approach, which subtracts from the value of the harvest only the paid costs. The alternative calculation, attributing the going wage to family labor, is used, however, as part of a later discussion of risk.

Returning now to the issue of tools, my discussions with Pasanos determined that they conceptualize the shovel or tobacco sprayer as an indivisible cost sustained in one year. They know that shovels *should* last four or five years, but they also know shovels *can* break after one year. Sheds can blow down, or tobacco contracts can be lost. Thus, in the following calculations, these fixed costs are deducted from profits in the same way as variable capital costs, because of uncertainty as to the accurate length of time over which to depreciate a tool, plus the simple fact that the cash to pay for it must come out of the family's budget in the year of purchase. In cases in which tobacco sheds were financed on credit, the yearly payment to the bank is counted as the cash cost.

Pasture as a land use option must be calculated slightly differently, but the methodology is as consistent as possible both with Pasanos' statements and with the calculations for the other options. Farmers who raise cattle say they calculate the annual return by subtracting cash costs from gross income. Income is derived either from the value of the calves sold or from the imputed increase in the value of the livestock, if none are sold in that year. Only cases in which animals were sold, however, were used in the figures below to avoid the bias of an estimated increase in value. Some households keep cows only for the sale of calves, but most also milk them, and I added the value of the milk to the value of calves in determining gross income each year. This methodology parallels that used by Pasanos themselves. One woman who had bought a cow said her calculations showed that after milking it for 15 months she had gotten back all of her investment solely in the value of the milk.

Although the expected productive life for cows is no more certain than the durability of a shovel, the expense is so much greater that it cannot be absorbed in one year as an indivisible cost. The majority of Pasanos use bank loans to finance some of their cattle purchases, and the quantity of credit for cattle production in Costa Rica has been considerable in recent years (see Meehan 1978 and Chapter 8 following). In spite of this pattern, however, no farmer discussed the original cost of breeding cattle in his cost-benefit analyses for pasture. There are two possible reasons for this. One is that, especially for the larger landholders with more than ten cows, some cows will have been purchased as adults, others will be cows reared from calves born on the farm, and still others will be cows reared from purchased calves. To unravel exactly how much capital has been invested (and foregone from the sale of calves) is too complex. The second reason is that the investment in cattle is seen as an investment in the whole farm enterprise, not specifically in any one piece of land on the farm. This interpretation is supported by the fact that some households own cattle that they keep on their neighbors' pasture, and pay rent.

Taken together, these reasons suggest that the value of "seed" cattle is an issue separate in the farmer's mind from his or her earnings per manzana of pasture. Nevertheless, for our purposes, this cost of breeding cattle must be included. The average cost of buying a cow in the year of study is ₡1,190, though this figure is probably high, given that a percentage of cows are raised from calves. It is estimated that the cow can be kept for a maximum of ten years, after which it will be sold as meat. This estimate is probably some-what high for the productive lifespan of a cow, but it tends in the opposite direction from the inaccuracies in the cost estimate. After ten years, the cow is sold for one-half to one-quarter of its purchase value, an average of ₡393, leaving an investment cost of ₡797 to be distributed over its ten-year life, ₡80/year/cow. Data from the interviews show that there is an average of .45 of an adult animal per manzana of pasture in Paso. Hence, ₡80 × .45 = ₡36 per year per manzana is the estimated capital cost of the "seed" cows. This estimate is combined with the other cash costs such as vaccinations and salt and subtracted from the gross profits of the pasture.

For the pasture calculations, labor is another problem. While farmers readily discussed the amount of labor necessary to weed pasture, to chop seedlings down, they flatly refused to discuss the amount of labor necessary to care for the animals. In contrast to their willingness to estimate all other labor inputs, they said the work with cattle is done is such tiny bits, on the way to a field or coming home at night, that it cannot be calculated. Therefore, based on my observations and farmers' discussions of the tasks involved, I have added one whole workday a month as an estimated labor cost. Together with the annual average of 2.62 peons per manzana for the weeding, then, the labor cost of pasture comes to 14.62 jornals per year.

Reliability

In any retrospective eliciting of production information, the question of reliability is important, and I used several methods to evaluate the Paso data. For labor figures, distortions inherent in farmer estimates are minimized by the small size of the tasks being estimated. For example, even with large plots of corn, weeding rarely takes more than two weeks. Also, the number of workers involved helps keep the total more accurate; corn weeding may require two weeks with one man working, or two weeks with three men working. Only with tobacco, where up to two months may be spent in one task, can the accuracy of Pasano estimates be doubted. But information from a government study brings strong confirmation. The Tobacco Defense Board did a detailed study of 69 tobacco producers in another area of Puriscal, recording each day after work the amount of time spent by all members of the family, prorating work of women and children to adult male equivalents. These figures were compiled and an average of jornals per manzana was calculated at 206 jornals. The average of Pasano estimates of tobacco labor is 220 jornals. This minor (7%) difference between observed and estimated annual labor input is significant validation of these estimates.

Figures on coffee and tobacco sales were obtained from the companies that purchase these crops, and I compared their records with Pasanos' statements. In the case of tobacco, for instance, 57% of the informants reported earnings exactly, while 93% reported correctly to within 5% of what the records showed. Purposeful distortions were rare in these production data, reflecting the "openness" of this open peasant community. Also, because Pasanos sell much of their produce directly by weight, they are more aware than many peasant agriculturalists might be of the exact measurements of their harvests. In addition, observations were made of all the agricultural tasks and samples of grain harvests were weighed. Both activities bore out the accuracy of farmers' estimates.

Of course, there are variations in each farmer's ability to estimate labor inputs and harvest. These variations in estimation fall on both sides, too low and too high. By averaging the Paso figures and using the statistics in the aggregate, this bias is overcome. In the estimates of harvest, a few cases were suspect—cases of very high harvests claimed by a poor household or very low yields claimed by a wealthy landholder. Where other means of validation suggested these harvest figures to be seriously distorted, three such cases were dropped. My observations also showed significant differences between families in the quality of work. In tying tobacco to poles, for instance, one family working all day tied 30 poles while another family with an equal number of workers tied over 100. These differences in intensity of labor are to be expected in a population and are not taken into account in this analysis.

A type I (*rancho*) style house. The palm-thatched roofs and walls of vertical poles are rapidly being replaced by type II and III houses in Paso.

A type II house. Note the combination of metal and tile roofs and the typical front porch construction.

A type III house, with tiled front porch and curtained, glassed windows.

The rugged mountains of the Puriscal area require farming on steep slopes. Note the two tobacco terraces in the picture and the pastureland in between.

A commercial street in Paso's nearby market town, Santiago de Puriscal.

The social and geographical center of Paso—the *pulpería* or general store with the coffee buying station to the left. Some men are playing cards after the day's work.

Doubling corn with machete prior to broadcasting beans.

Classification of tobacco into "rolls" in the tobacco shed; the whole family works together at this task.

Costa Rican cattle; note the interbreeding of native cattle with Brahma strains from India.

Tobacco drying shed, protected from the occasional dry-season rain by palm thatch, metal sheets and plastic.

PROFIT

Pasanos discuss profits perhaps more than any other factor in making land use decisions. Though they may phrase it in general terms such as "this crop pays" or "gives you something," they generally say that tobacco is the most profitable, with coffee also highly valued.

Tobacco gives me the most.

I started tobacco to make more money than with other crops.

Coffee is good money and not risky.

I want to start tobacco because I can't make money with beans anymore.

The results of the profit calculations discussed above are presented in Table 5.1. These etic figures support the emic assertion that "tobacco gives me more"—its profit level is three times that of any other crop. Coffee is also quite profitable compared to pasture, but corn and beans are higher still, even though no Pasano indicated traditional grains to be a profitable option. These profit figures indicate why tobacco has grown in importance in recent years as contracts have become available. But the shift to pasture, while shown in Chapter 4 to reflect rising beef prices, is shown by Table 5.1 to be uncompetitive with the other three land uses in terms of profit per land unit. With these profit comparisons in mind, the other criteria of utility used by Pasanos can now be explored.

TABLE 5.1. RETURNS TO LAND: NET PROFITS PER MANZANA PER YEAR (IN COLONES)

Traditional grains	Tobacco with grains	Coffee	Cattle
₡1,242	₡3,707 Tobacco only: ₡2,556 Grains only: ₡1,151	₡923	₡340

NOTE: 1 manzana equals .69 hectares; 1.7 acres.
        1 colon equals $.12 U.S.

LABOR REQUIREMENTS

Pasanos discuss the labor input of each option:

Tobacco gives the most money, but coffee only requires fertilizing and harvesting, easy work. Corn is harder work.

Beef is the best—cattle grow themselves.

Tobacco is best if you have a peon to help—it's so much work!

Obviously, Pasanos disagree as to the appropriate decision to be made based on the amount of work required for each crop, but labor input is clearly a factor in their decisions. Pasanos evaluate not only the quantity but the quality of work involved for each crop. As discussed in Chapter 4, tobacco terracing is the most arduous of all agricultural labor. The other tasks of tobacco are not characterized as significantly more difficult than the tasks for most other crops. Coffee requires most of its labor in harvesting, which is easy, and the little labor required to maintain cattle is considered easy work, too.

The production data gathered for profits illuminate the quantity of labor required for each option. These averages are shown in Table 5.2 and have also been weighted. (See Table 5.2.) These figures show that the drudgery involved in producing a manzana of tobacco is more than four times the work needed for a manzana of traditional grains. Given that some of the labor is spent in terracing, tobacco's labor costs are very high. Grains and coffee are quite similar in labor investment—72 jornals and 85 jornals respectively. As for pasture, informants' statements that "beef is the best" are supported with the very low 15 workdays per year requirement.

TABLE 5.2. LABOR REQUIREMENTS: JORNALS PER MANZANA PER YEAR

Traditional grains	Tobacco with grains	Coffee	Cattle
72	305	85	15
	Tobacco only: 220		
	Grains only: 85		

NOTE: One jornal equals six hours.

In terms of drudgery, then, tobacco seems an unwelcome choice, while pasture is an easy option, the reverse of their profit utilities. Coffee and corn seem about equal in amount of work and profit, yet informants value coffee highly and have increased its plantings, while acreage in corn drops continually.

RETURNS TO LABOR

We can now explore the Chayanovian approach that farmers make decisions based on returns to labor. Combining the two preceding measures and dividing the profit per manzana by the number of jornals of labor required gives a calculation of net profit per jornal. (See Table 5.3.) Table 5.3 shows strong support for the interpretation that land scarcity has shifted decision making away from returns to labor. If Pasanos wished to invest their labor where it would bring the greatest yield, they would continue their traditional grain production rather than shifting out of grains into tobacco, pasture,

TABLE 5.3.   RETURNS TO LABOR: NET PROFIT PER JORNAL (IN COLONES)

Traditional grains	Tobacco and grains	Coffee	Cattle
₡17.25	₡12.15 Tobacco only: ₡11.62 Grains only: ₡13.54	₡10.86	₡22.67

and coffee. This table can be used as evidence that although returns to labor may be a factor in Paso decisions, other criteria have higher priority.

CAPITAL REQUIREMENTS

Tobacco is characterized by a number of Pasanos as "too expensive," while the capital costs of other options are not discussed. Production data, weighted and averaged, show the cash costs per manzana. (See Table 5.4.) Table 5.4 shows average cash costs for the four crop options. Beans, since they require no fertilizer or special tools and are rarely planted with hired labor, have no cash costs for almost all households. Cattle are clearly the cheapest, in spite of the estimated cost of the "seed" cows. They require mainly salt, medicines, and sometimes peons to maintain the pasture. Cash costs for corn, tobacco, and coffee come primarily from fertilizer purchases, but also include hired labor, tobacco sheds, sprayers, etc. The average cash cost for coffee is inflated somewhat by a few large landholders who hire a great deal of help in the harvest. The majority of Pasano households hire some help but pick most of their coffee themselves and therefore do not sustain such high costs. These figures support Pasanos' statements that tobacco is much more costly to produce than any of the other options. In the past, when bank loans for small farmers were rare or unavailable, these capital inputs deterred some farmers. Today, however, the availability of credit makes these costs surmountable by virtually any household (see Chapter 8). Their significance lies, rather, in the size of the risk involved in investing nearly ₡400 in a manzana of coffee or ₡970 in a manzana of tobacco and grains, versus only ₡63 in a manzana of pasture for cattle.

TABLE 5.4.   CAPITAL REQUIREMENTS: CASH COSTS PER MANZANA
             PER YEAR (IN COLONES)

Traditional grains	Tobacco and grains	Coffee	Cattle
₡131	₡970 Tobacco only: ₡803 Grains only: ₡167	₡388	₡63

RETURNS TO CAPITAL

Combining this last information with the overall profits per manzana, the return per colon invested in the four crop options can be computed. (See Table 5.5.) Table 5.5 shows that the traditional production of corn and beans gives the highest return for the relatively small amount of capital that must be invested. Cattle are the next most profitable use of capital, mainly because the amount of capital required per manzana is quite small. Tobacco comes out third, and the grains with it are much more profitable than the tobacco alone. Coffee, surprisingly, gives the lowest return, but this figure reflects the high labor costs sustained by some large landholders while comparing that high cost with the average profit for the whole community.

TABLE 5.5.   RETURNS TO CAPITAL: NET PROFIT PER COLON INVESTED

Traditional grains	Tobacco with grains	Coffee	Cattle
₡9.48	₡3.82 Tobacco only: ₡3.18 Grains only: ₡6.89	₡2.38	₡5.40

YIELDS

Before turning to a discussion of risk, the two forms of grain production in Paso need to be compared. Pasanos say:

> Tobacco is a bad crop because they can downgrade your harvest and pay you nothing for it and you come out scarcely making as much as a peon, but there is no other way to get good corn and beans. The grains are the essential part of tobacco.

> My [traditional] corn and beans weren't doing well—especially with the *babosa* [a slug], and the tobacco coop came along, so I started tobacco.

Table 5.6 compares the average community-wide production of corn and beans by the two methods. The yields of corn show that corn rotated with tobacco produces significantly more than traditional corn. Bean figures show the reverse. Here, however, a caveat must be entered concerning the tradi-

TABLE 5.6.   GRAIN YIELDS PER MANZANA PER YEAR

Crop	Traditional	In rotation with tobacco
Corn (in cajuelas[a])	73	111
Beans (in quintals[a])	6.5	4.6

[a] 1 cajuela = approximately 33 pounds; 1 quintal = 100 pounds.

tional bean figures. In recent years Pasanos have suffered badly from pests that attack broadcast beans, and now there is only one area of the mountain (an area primarily owned by large landholders) where broadcast beans can still be grown successfully. These high bean yields (and some of the high profit shown for traditional grains) reflect only the best land, and it would not be possible for the majority of the community to obtain such a high yield.

At first glance, the higher production of beans in traditional fields seems to balance off the higher corn yields in tobacco terraces, but this is not so. In terms of cash value, the extra corn is worth more than the extra beans. Also, in terms of calories, the higher corn yield in rotation with tobacco produces more food value than the higher bean yield in traditional plots. Since most Pasanos cannot get good bean yields in traditional plots, their shift to terraced grains is not a sacrifice of protein for increased calories. The grains planted in tobacco terraces are, therefore, clearly more productive than grains in traditional fields, especially when the problems with bean production are taken into account.

RISK

Risk and the probability of good harvests are discussed frequently by Pasanos:

A good year of beans gives you more than tobacco, but a good year is rare now.

I'm tired of tobacco because it has so many problems, especially the wind.

It is hard to lose a crop of coffee; beans are very risky. Corn has risks but the main problem is its low price.

A number of recent works have discussed the issue of risk, uncertainty, and farmers' decisions at length (Berry 1980; Cancian 1979, 1980; Chibnik 1981; Ortiz 1980; Roumasset 1979; Roumasset, Boussard, and Singh 1979). Berry has challenged Knight's much-used distinction between risk and uncertainty (*risk* defined as measurable, quantifiable probability of loss and *uncertainty* as the unknown probability) by saying that "probabilities of future events are never 'known' with complete certainty" (Berry 1980:325). Ortiz's work with the Paez of Colombia has shown that farmers are not able to discuss a weighted set of probable outcomes on which decisions are based but rather formulate some expectations as to price and yield based on a range of recent events. Thus, the classic view that farmers who are familiar with a crop know "the odds for and against" a certain outcome is questioned by these writers.

Evidence from Paso conforms to this perspective. Pasanos are aware of much variation in harvests, price, and profits but do not discuss odds for bad weather or good years. When forced, some can grudgingly guess at the frequency of bad yields, but most prefer to discuss instead the general riskiness of certain options and then what happened in particular disaster years. Since long-term profit or yield figures were not available from any other source, I measured Pasanos' general statements on the comparative riskiness of certain crops by comparing the proportion of unacceptable outcomes for the community as a whole, based on the profit calculations discussed previously. This variance within one year will serve as a flawed but still useful proxy for a longer sequence of outcomes, especially since Ortiz has found that expectations of outcomes combine both prices and yields and conform most closely to recent past history (Ortiz 1980:188).

To measure the variance in "acceptable outcomes" for one year, the traditional cost-benefit calculations were used, valuing the unpaid family labor at ₵6 for six hours of general agricultural work and ₵10 for six hours of tobacco terracing. These attributed costs, together with all cash costs, are subtracted from the gross value of the harvest. Although farmers do not necessarily reject a crop option that pays them less than the going wage for their labor, this criterion reveals the probability of negative profit outcomes in the community in one year, the best measure available to assess risk. The year under study was said to be an average year; harvests were neither exceptionally high nor exceptionally low.

Table 5.7 shows that tobacco by itself is, indeed, the riskiest crop of all—one-third of all households producing tobacco did not make enough from its sale to repay themselves for their labor. Traditional grain production is also quite risky, as one informant pointed out above, but grains in rotation with tobacco are strikingly less so; only 9% of the outcomes were negative. Thus, in terms of risk, Pasanos' willingness to adopt tobacco can be seen as accepting slightly higher risk for one part of the rotation and much lower risk for the other. Since tobacco occupies two-thirds of the year in the field and the grains the remaining third, weighting and averaging these two figures gives a risk figure of 25% for the tobacco/grains rotation as a whole. As Wharton predicts, "The variability in the expected results with the new technology" is measured against "the minimum subsistence standard and the variability experienced with current technology" (Wharton 1971:170).

TABLE 5.7.   PROBABILITY OF NEGATIVE OUTCOMES—1972

Traditional grains	Tobacco	Grains with tobacco	Coffee	Pasture
N = 44	N = 33	N = 19	N = 41	N = 14
24%	33%	9%	12%	0%

Thus, the high overall risk of tobacco (25%) is evaluated more favorably, given the equally high risk of traditional corn and beans (24%).

Coffee ranks third but is very low in risk, and most households in this 12% have new coffee that is not fully productive or old coffee that is past producing well. As one farmer said, "It's hard to lose a crop of coffee." Pasture is shown to have the lowest risk of all; no households failed to get a fair return on their labor in this land use. In sum, although this measure is a somewhat arbitrary calculation of risk, the results in Table 5.7 conform strikingly to farmers' statements and to my own subjective assessment of the riskiness of the four options.

SUMMARY OF THE CROP OPTIONS

Table 5.8 summarizes the preceding evaluation of the four crop options based on Pasanos' statements and etic measures. Tobacco is shown to give the highest return to land and also to give the highest yields of corn and beans (also at lower risk than traditional corn and beans). Pasture gives the highest returns to labor with the lowest risk and is second in returns to capital, making it a safe and lucrative investment for large landholders. Traditional grain production gives the highest returns to capital and gives substantial returns to land and labor as well. Those farmers who continue this crop option are not acting from "peasant conservatism" but rather from sound economic planning. As noted, many Pasanos can no longer plant traditional corn and beans successfully and so have shifted their land to other uses, leaving these traditional grain figures quite high, since they reflect only the most productive fields. Coffee emerges as a safe and otherwise undistinguished land use.

THE EVALUATION OF TOBACCO

We can now return to the question with which this chapter began: why have Pasanos become so interested in tobacco? The characteristics of the crop are clear; tobacco is both an attractive and unattractive option. It gives a very high profit, with very high risk. Its harvest is followed by grains, which are the least risky of any of the agricultural options except for pasture, and gives a higher productivity than grains in traditional plots. Tobacco terracing is backbreaking, arduous work, but it actually improves the soil's fertility. The hours needed for tobacco are greatly in excess of the requirements for other crops, and the profit per jornal in tobacco is much lower than for grains. The capital investment required for tobacco is also very high, and this capital must often be borrowed, with all the problems and risk that credit entails. The sale of tobacco is problematic, and supervision during the growing season is sometimes onerous.

TABLE 5.8.   RANK ORDER OF CROP OPTIONS, BY FIVE CRITERIA

Criterion	Traditional grains	Tobacco and grains	Coffee	Cattle
Returns per				
land unit	2	1	3	4
labor unit	2	3	4	1
capital unit	1	3	4	2
Risk preference	3[a]	3[a]	2	1
Yields	2	1	(not applicable)	

[a]Tobacco/grains and traditional grains are shown here as tied for third place (see text).

In spite of these drawbacks, the scarcity of land in the community makes tobacco attractive. The primary criterion of decision making for the majority of households must be the returns to land, and here tobacco far exceeds its nearest competitor. Tobacco allows for three crops from the same field each year, and the good grain production helps to assure the family's consumption needs, while the sale of tobacco helps with the family's cash needs. This option is then doubly adaptive in a situation of population growth and declining access to rented land, and Pasanos accept both the lower returns to labor and capital and the high risk in choosing to increase tobacco production. Farmers' reluctance to count their unpaid labor costs is now better understood; in the context of land shortage, the traditional importance of returns to labor must be reduced. Farmers do not like to remind themselves of the high costs of putting returns to land first: "We don't come out making anything then."

6

Differential Access to Land and the Pasture Decision

As outlined in Chapter 5, measures of population pressure must be modified in a state-level society to take into account unequal access to productive resources. This chapter pursues the implications of that stratification in land resources: that each farmer's decision-making process is affected by the household's access to land. Through an analysis of land use choices by stratum, the process by which large landholders choose pasture can be understood. In addition, the variations in Pasanos' statements about the four crop options also become clear.

LAND USE BY STRATUM

Cancian (1972, 1979), Durham (1979), DeWalt (1975, 1979a), Finkler (1980), Takahashi (1970), and Acheson (1980), among others, have shown in recent research that in rural communities stratification and differential access to land profoundly affect agricultural decisions. Not only affected are the crops chosen and the innovations adopted, but also the amount of land planted to each crop chosen. The social strata discussed in Chapter 3 were analyzed for land use patterns. Table 6.1 shows the typical crop mixes for each stratum: the mean number of manzanas dedicated to each crop is shown for each stratum, and the mean number of manzanas for all households in that stratum who actually planted that crop is shown in parentheses. The latter

TABLE 6.1. LAND USE BY STRATUM: AVERAGE NUMBER
            OF MANZANAS PER HOUSEHOLD

Stratum	Traditional grains	Tobacco with grains	Total grains	Coffee	Pasture
Landless (N = 13)	.8 (.9)	.3 (.9)	1.1	—	—
Heirs (N = 8)	1.4 (1.6)	.5 (.8)	1.8	.3 (.6)	—
Small (N = 23)	.7 (1.0)	.6 (1.1)	1.3	.8 (.8)	.5 (1.3)
Medium (N = 17)	1.0 (1.6)	1.5 (2.2)	2.4	1.9 (2.0)	8.3 (10.1)
Large (N = 8)	2.0 (2.3)	[a](4.0)	2.3	2.0 (2.6)	58.4 (58.4)

[a]One family has recently purchased enough land to put them in the "large landholder"
category. They currently allocate 4.0 manzanas to tobacco. This is the only cell of the table
where *n* is less than 3.

figure gives a less distorted idea of the average plot size for any one crop.
Minor land uses such as rice, fallow, and forest are omitted from the table.
Table 6.1 supports two general points about land use choices in Paso. First,
most strata have a crop mix, thereby benefiting from the different characteristics of the various options. Labor requirements are spread out over the year
("I started tobacco because it is work in an off-season from other crops").
Risk is spread out as well (Johnson 1971b:145; Ortiz 1976:16), both because
different crops are susceptible to different diseases, climatic problems, and
harvest failures and because they are subject to different external marketing
and price fluctuations.

Second, land use varies by stratum both in terms of what is planted and
how much. Table 6.1 shows that grains are a high priority for all strata; all
households plant corn and beans, either with tobacco or by the traditional
method. Tobacco is planted commonly by all strata except the large landholders, while coffee is planted by all except the landless. Pasture is re-
stricted to owners of land—no heirs or landless use their rented land for
cattle. The most complex land uses are chosen by small and medium
landholders, who tend to have a four-part mix.

A basic pattern of agriculture for Paso seems clear. All households plant
traditional corn and beans. As Table 6.1 shows, families with access to larger
amounts of land all plant coffee. After planting a stand of coffee, additional
landholdings will be planted to pasture. It is interesting that even large
landholders, who could dedicate themselves entirely to cattle ranching,
continue to plant a mix of traditional grains and coffee. This pattern is
evidence of the extent to which the large landholders are still farmers and not
a separate social class of "cattlemen on their horses." I attempted to develop a
similar table, using household size. No clear patterns emerged; small,
medium, and large families plant a similarly wide range of crop mixes. Since
access to capital closely parallels and is contingent upon access to land, these

patterns by stratum show that access to land is the primary determinant of the kinds of land uses Pasanos choose.

Within these general patterns of crop choice, we can turn to the question of the *quantity* of land planted. What are the basic determinants of the amount of land planted to each crop? Since corn and beans are the main subsistence crops grown in Paso, Chayanov (and others; see Chibnik 1974) expect that the amount of grain land planted will vary with the *size of the household* (Chayanov 1966:67). This hypothesis was tested using a contingency coefficient (which can vary from 0 to .816 in this case); the relationship between these two variables is .24 (not significant). The size of the household is not related to the amount of land planted to coffee or pasture, either. It is related, however, to the amount of land planted to tobacco, and this pattern will be discussed in the next chapter.

Another factor possibly influencing the amount of land planted to each crop is the *age of the household*. As discussed in Chapter 3, landholdings increase over the family cycle, and as landholdings increase, so do the amounts of land in pasture. The relationship between amount of pasture and the age of the household is statistically significant at the .01 level (C = .44). As might be expected, no other crop choice shows this pattern.

Amount of land owned, however, turns out to be closely related to the amount of land planted to each crop. For traditional corn and beans, coffee, and pasture, the amount of land dedicated to the crop varies with the amount of land owned at the .01 level. Tobacco shows a similar pattern, but at the .1 level of significance; this lower correlation reflects the large landholders' avoidance of tobacco production.

Thus, the analysis of land use choices in Paso according to access to land shows not only that certain crop mixes are characteristic of certain strata but also that the amounts of land allocated to certain crops vary by landownership as well. These differences, not surprisingly, are reflected in Pasanos' statements.

Patterns in Pasanos' Statements

In the quotations in Chapter 5, there is some disagreement not on the criteria of choice themselves but on how those criteria are to be used; different criteria for the same crop are stressed by different speakers. By linking the speaker's stratum to the character of the comments made both to me and in casual conversations among other farmers, some patterns emerge that closely parallel the findings of Table 6.1.

While all farmers discuss the many drawbacks to tobacco—high risk, high cost, high labor investment—a proportion of the community concludes that tobacco is still worth the trouble. "Tobacco is a lot of work, but it gives you something for it; corn and beans [traditional] don't anymore. And the corn and beans in the terraces do well." Informants who feel this way are spread more or less equally among all the four strata, except large land-holders. Large landholders do not see tobacco as "worth the trouble," a conclusion that is appropriate to their situation of access to plentiful land. One large landholder says, "I have never liked tobacco and never wanted to grow it. I don't like the work; it's boring."

Another cluster of statements admits that tobacco is very profitable, but its risk and labor input are too high. These informants maintain that coffee is a better crop. "Tobacco is too much work; coffee is easy work and always gives a profit." This group consists only of a few small and medium land-holders who do not have enough workers to undertake tobacco; the reasons for their rejection of tobacco will be discussed more in the following chapter. No landless, heirs, or large landholders expressed the judgment that coffee is the crop for them.

Predictably, statements evaluating pasture as the most profitable land use come from large landholders and the heirs of large holders. One woman says, "We don't want to plant more coffee because pasture gives more than coffee. Cattle are cheaper and easier." While it has been shown that profits per manzana of pasture are actually very low, these emic evaluations see pasture as "most profitable." No landless, small, or medium landholders evaluate the pasture option this way.

It is fascinating to note that throughout the year of research, four households indicated that their favorite crops were either rice or traditional forest root crops. All four of these households are extremely poor, have adopted few or no innovations such as fertilizer, and generally lack sufficient land to rent. The family that prefers root crops has no man to do any agricultural work, and though the women try to plant corn and beans and maintain a stand of coffee, they say they prefer root crops to other types of agriculture. Since root crops require only planting and harvesting, the women's opinion is understandable. Rice, the preferred crop of the other three farmers, can no longer be grown in the community. It is significant that preferences for the "old" crop options are expressed by the least successful and least respected farmers in the community.

THE PASTURE DECISION

This analysis of land use by stratum lays the foundation for understanding the increase in land dedicated to pasture in Paso. Though the profits per land unit were shown to be very low, some farmers evaluate pasture as very profitable. Their calculation can now be clarified, given the information in Table 6.1 that most pasture is in the hands of large landholders.

With large amounts of land available, the large farmer is no longer bound by the constraints of the rest of the community. Instead, labor becomes the scarce factor that affects his crop choices. The pool of wage laborers, however, has its limits, and there are perceived shortages at certain times of the year. The large landholder cannot choose, for instance, to put all of his 65 manzanas into a labor-intensive crop like coffee. It would even be difficult to find laborers for five manzanas of traditional grains, especially since the phases of the growing season require certain tasks to be done promptly, or harvests will be affected. These labor constraints favor the use of pasture, which requires the least labor input per manzana. Given that beef prices have gone up sharply in recent years, the large landholder who wishes to maximize the income from his farm has found that taking land out of fallow or forest is a profitable change. To explore this calculation, a hypothetical budget from a large farm ("Fulano's Farm") can be constructed from Table 6.1 and from the profit averages in Table 5.1. Fulano thus has a total of 64.7 manzanas allocated according to the community averages: 2 manzanas for traditional grains, 2 for coffee, and 58.4 for pasture. Though not included in Table 6.1, the average large landholder in Paso also has an additional 1.3 manzanas in fallow land and 1.0 in forest, and these are part of the 64.7 total. The profits from this hypothetical allocation of land are shown in Table 6.2.

TABLE 6.2. FULANO'S FARM: CURRENT LAND USE

Land use	Number of manzanas	Profit per manzana	Total per year
Grains	2.0	₡1,242	₡ 2,484.00
Coffee	2.0	923	1,846.00
Pasture	58.4	340	19,856.00
Fallow	1.3	0	0
Forest	1.0	0	0
	64.7		₡24,186.00

To understand how Fulano has come to prefer pasture over traditional grains, this hypothetical farm budget can be redone to produce the maximum under the traditional land uses. We will assume that the allocations in coffee and forest are optimal. The coffee brings high profit with low risk and utilizes family labor at harvest time, so its harvest is not dependent on peons. Forest products are necessary, not only for firewood for cooking but also for construction and other farm tasks. To calculate the quantity of land available for grains, it can be assumed that five years of fallow are necessary to maintain soil fertility in slash-and-burn fields in Paso (see Chapter 1). Fulano has 58.4 manzanas of pasture, 1.3 of fallow, and 2 of grains available for traditional grains production: 61.7 manzanas total. One-sixth of this land can be in use at any one time, which would allow him 10.3 manzanas of grains. This figure is greater than the 5 manzanas mentioned above as a probable ceiling on the available labor pool. Therefore, it will be assumed that Fulano

TABLE 6.3.   FULANO'S FARM: TRADITIONAL GRAINS INSTEAD OF PASTURE

Land use	Number of manzanas	Profit per manzana	Total per year
Grains	5	₡1,242	₡ 6,210.00
Beans	5.3	435	2,305.50
Coffee	2	923	1,846.00
Fallow	51.4	0	0
Forest	1.0	0	0
	64.7		₡10,361.50

can produce a maximum of 5 manzanas of traditional grains and will do the remaining 5.3 manzanas in broadcast beans. The profit figure used here for beans assumes that all labor is paid. His profit from his 64.7 manzanas is only ₡10,361.50 by doing the maximum possible with traditional grains. In pasture, he earns ₡24,186.00.

It can be suggested, however, that if his land were properly rotated with fallow and planted to beans, Fulano would have more fertile soil than the present average in Paso, though this average is already recognized to be skewed somewhat high. In the other direction, growing larger stands of corn and beans may produce somewhat lower yields, since the fields are not as carefully watched and since peons may not work as hard as Fulano himself would. It may be arbitrarily assumed here that his profit might be 25% higher, given hired labor and good fallow periods. The profit per manzana for corn and beans then becomes ₡1,552.50, and the profit for beans alone is ₡543.75 (see Table 6.4).

The large landholder's total income from his land in grains is *still* less than the ₡24,186 he can make with pasture. And taking into account the very low risk with cattle, the larger farmer is much safer raising cattle than he is raising grains with their much higher risk. Furthermore, he avoids the problems of finding and supervising the peons to work his ten manzanas of corn and beans. The high price of beef has shifted his optimal land use, and large landholders have changed their land uses accordingly.

TABLE 6.4.   FULANO'S FARM: HIGHLY PRODUCTIVE GRAINS

Land use	Number of manzanas	Profit per manzana	Total per year
Grains	5	₡1,552.50	₡ 7,762.50
Beans	5.3	543.75	2,881.88
Coffee	2	923.00	1,846.00
	12.3		₡12,490.38

The land use evaluations of large farmers, both emic and etic, maximize the overall returns to the farm, not returns per manzana. Given the estimated labor requirements discussed in Chapter 5, they are not minimizing labor costs, however. Two manzanas of coffee, 5 of grains, and 5.3 of beans would require 678.4 jornals. Two of grains, 2 of coffee and 58.4 of pasture would require 1,190 jornals. This latter figure includes 876 jornals spent in cattle production and obviously does not take into account the labor economies of scale in having many cattle. Returns to labor in the current land use are, however, higher than for the hypothetical grains combination—₡18.94 versus ₡15.27 per jornal. Minimization of risk is also clearly included in the decision-making process, as are returns to capital. There is no evidence from large landholders that one or another of these criteria is primary; rather, they are all part of the decision that forest and fallow are less desirable uses of large tracts of land now that the price of beef has risen much more than the price of corn and beans.

The consequences of this shift to pasture are in some ways opposite from those of the shift to tobacco. Pasture increases the distance between the large landowners and the landless and makes upward mobility less likely for those who do not inherit land. Pasture also increases the pressure of population on the land by removing large areas from fallow and making them unavailable for rental. The effects on Costa Rica as a whole are serious as well; soil erosion and water shortages from decreased water retention in the dry season make pasture an undesirable land use for such shallow, tropical soils. The decisions of large landholders in rural areas seem to be motivated by short-run considerations and may not even seriously consider long-run ecological results. Given the current situation in which beef prices (and thereby the future of the pasture option) are determined outside Costa Rica, long-run trends may, in fact, *not* be a sound basis for farmers' land use decisions.

7

Household Resources and the Prediction of Land Use Choices

In this chapter, we turn to the issue of *prediction*. Although it has been established that each stratum has a characteristic land use pattern, how much variation is there within the stratum? How can we predict what any one household will plant? Having begun with a community-level analysis, we have moved to the patterns within the five strata and can now turn to the individual household level, and the behavior of "Tom, Dick, and Harry." Two theories of land use—Chayanov's and Boserup's—will be tested to see if their predictions are useful in understanding the decision to plant tobacco in Paso. The results of this analysis will then allow us to develop a flow chart of crop decisions in Paso, combining the factors of access to land and household labor resources. The flow chart successfully predicts up to 89% of farmers' choices and illustrates a number of important points about agricultural decisions on the household level.

CHAYANOV'S APPROACH

In the theory of peasant economy developed by the Russian economist Chayanov (1966), the prime mover of household decisions is the consumer-worker ratio, that is, the ratio between the workers in a family and the number of mouths that they must feed. One important consideration in using Chayanov's theory is that he assumes land is abundant and can be purchased or rented freely. Under this assumption, a family that finds itself pressed to feed an increasing number of children will expand the farm. The labor required to feed the children will also increase, and Chayanov sees the consumer-worker ratio as the determinant of the farmer's level of "self-exploitation." If land is abundant, then the technological level of land use need not change; self-exploitation would mean only that each worker works more land. If, however, this assumption does not hold and land is scarce, Chayanov notes that farmers will have to work the land more and more intensively in order to feed their families (Chayanov 1966:8). The rest of Chayanov's perspective, however, remains the same in a situation of scarce land: households with a higher consumer-worker ratio would have more need to intensify. Thus, for the tobacco decision, households with a higher consumer-worker ratio would be more likely to adopt tobacco. Likewise, as children grow up and help with the work or leave home, the ratio will drop and households should need tobacco less. How well do Pasanos conform to these expectations?

In order to apply Chayanov's theories to Paso, several changes have to be made. When computing "workers," Chayanov included all family members 19 and over as full workers (0.9 in value), while those aged 14 to 18 were counted as 0.7. He does not discuss the extent to which Russian women participate in agricultural work or in agricultural decisions, but they are included equally with men as workers. In Paso, however, women contribute only a small amount of their yearly labor to agricultural tasks. Field work is clearly considered men's work, while the women's domain lies within or near the house. Thus, it seems more appropriate in the analysis of agricultural decisions for Paso to calculate the number of workers as the number of adult males, not the total number of adults. This pattern will be discussed more in another section. Also, boys help their fathers from the time they finish school, usually at age 13. While they do not equal the strength of their fathers at that age, they can and often do equal their fathers' hours of field labor. Since comparative measurements of labor intensity are not available, all males over 13 and under 60 are classified as adult male workers.

To compute the number of "consumers" in each household, I used Schultz's method, whereby children under 9 years of age are valued at .50 of an adult, children 10 to 15 are value at .75, and all those 16 and over are valued as one whole adult (Schultz 1945:114). This formula gives the total number of adult equivalents (AE) in each household. The consumer-worker ratio for Paso is therefore the number of adult equivalents in each household divided by the number of adult male workers (AE/males). Some households have no adult male workers and are separated from the others in the following analysis.

Tables 7.1 and 7.2 present the relationship between the consumer-worker ratio in Paso and the decision to grow tobacco. Table 7.1 shows the

TABLE 7.1.   TOBACCO PRODUCERS AND THE MEAN ADULT EQUIVALENT/
MALES RATIO BY STRATUM

Stratum	Tobacco producers	N	Not tobacco producers	N	Number of households with no males
Landless	4.35	5	3.14	8	0
Heirs	4.11	6	2.50	2	0
Small	4.25	13	3.30	5	5
Medium	3.51	13	2.40	2	2
Large	2.92	1	4.66	6	1
Whole community	3.95	38	3.49	23	8

average number of persons supported by each male between 13 and 60, by stratum. For four of the five strata, households who plant tobacco have a higher ratio of consumers to workers than do those who do not plant tobacco. Leaving out the 8 households with no male between 13 and 60, there remain 61 households in the community, of which 38 plant tobacco and 23 do not. The average adult equivalent/males ratio is higher for the tobacco growers (3.95) than for the others (3.49), but this difference is not statistically significant at even the .1 level (using a paired t test).

Table 7.1 seems to suggest that Chayanov's consumer-worker ratio holds for four of the five strata in Paso, and thus the relationship between the AE/males ratio and tobacco production may be more significant if the large landowning households are not included. Omitting these eight households from the calculation, the mean AE/males ratio for tobacco producers becomes 3.98 and the mean for non–tobacco producers is 3.08. The difference between these means is statistically significant at the .01 level, using a paired t test. The consumer-worker ratio thus is shown to vary quite closely with the tobacco decision for the four less wealthy strata, though there is much more variation in the relationship when large landholders are included.

Using this evidence, we can now explore whether the Chayanovian AE/males ratio will neatly divide the community into tobacco growers and nongrowers. Table 7.2 omits the large landholders and divides the remaining households into two equal groups, using the midpoint AE/males ratio of 3.33. We would expect to find most of the families choosing tobacco on the

TABLE 7.2.   TOBACCO PRODUCERS BY ADULT EQUIVALENT/
MALES RATIO, NUMBER OF HOUSEHOLDS
(LARGE LANDHOLDERS OMITTED)

Producers	Ratio under 3.33	Ratio 3.33 and over	Totals
Tobacco producers	15	22	37
Not tobacco producers	11	5	16
Totals	26	27	53

right-hand side of the table and most of the abstainers on the left. The figures in Table 7.2 do not conform neatly to this pattern. Of the 26 families with ratios lower than 3.33, 15 plant tobacco and 11 do not. Clearly, tobacco production is not always forced on a family by consumer demand. The 27 families with higher ratios, however, are more uniformly forced into intensification: 22 grow tobacco and only 5 do not. While this latter difference is significant at the .01 level, the Chayanovian ratio is not useful to predict who will grow tobacco—of the 37 households that grow tobacco, we see that 15 of them have lower consumer-worker demand and 22 have higher. Even for the lowest consumer-worker ratios in the community, Chayanov's expectations are not upheld. Of the 8 households with a ratio of 2.0 or under, 4 grow tobacco and 4 do not.

Tables 7.1 and 7.2 show that intensification of land use, as seen here in the decision to grow tobacco, is more likely to be found in the families with a high worker-consumer ratio, when large landholders are omitted. But for these four groups or for the community as a whole, the Chayanovian ratio alone is not a strong predictor of who will grow tobacco and who will not.

We can now explore the assertion that the number of adult males in a family is a more appropriate measure in Paso of the number of workers than is the number of adults of both sexes. Table 7.3 compares the tobacco-producing status of households with and without a man between 13 and 60. The figures show that as long as there is one man between 13 and 60 in the household, some households will choose to grow tobacco. None of the eight households with no male worker grows tobacco. In four of these families, there are women workers under 60, yet still no household undertakes tobacco without a man who can do the heavy labor involved. These figures suggest that the presence of a grown man may allow a household to do tobacco but cannot determine whether or not it will choose to do so. Chayanov's theory of the worker-consumer ratio has been shown to be unable to predict which households grow tobacco, but labor resources are shown here to have a clear impact on the tobacco decision.

BOSERUP'S APPROACH

The second theoretical perspective that may help explain why some households plant tobacco while others do not is articulated by Boserup (1965), Basehart (1973), Brookfield and Hart (1971), Clarke (1966), Dumond (1965), Harner (1970), Netting (1968, 1974), Geertz (1963), and Spooner (1972) among others. Like Chayanov's, this perspective expects that when land becomes scarce, labor will be invested more heavily in order to produce greater yields. While Chayanov expects that the "self-exploitation" of the farmer is temporary and a part of the natural household cycle, this second perspective (which for convenience I shall call Boserup's) sees labor inten-

TABLE 7.3. TOBACCO PRODUCERS AND MALE WORKERS,
            NUMBER OF HOUSEHOLDS

Producers	No males 13–60	1 or more males 13–60	Totals
Tobacco producers	0	38	38
Not tobacco producers	8	23	31
Totals	8	61	69

NOTE: Chi^2 is significant at the .01 level.

sification as an ecological imperative based on a new human-land ratio. As long as the number of people continues to rise on a given amount of land, or the amount of land available continues to fall for a constant number of people, the need for increasing labor investment is expected to continue as well. Thus, Chayanov's theory is based on a relatively stable human-land ratio in the community and focuses on the fluctuations in population and resources at the household level. Boserup, on the other hand, is looking at the community level and does not postulate how each household will respond to overall population pressure.

In a stratified community where access to land is unequal, Boserup's approach can be translated to the household level by suggesting that small farmers will feel population pressure sooner than larger ones, using access to land in a system of private property as the boundary of agricultural decisions. A second important variable will be the size of the family, since that will determine consumption needs. A third issue that arises both in Boserup's and Chayanov's theories is the standard of living; both perspectives assume that production is primarily for subsistence. Neither author measures labor intensification arising from a desire to increase consumption. This point will be discussed more in Chapter 9.

To apply Boserup's ideas of land scarcity and population pressure to the tobacco decision in Paso, we can compare the ratios of households' land resources (either owned or rented land, measured in manzanas) to their subsistence needs (measured in adult equivalents). If farmers are intensifying their land use because of population pressure, the lower the land/AE ratio the more likely they should be to plant tobacco. Tables 7.4 and 7.5 present these data.

Table 7.4 shows some support for this use of Boserup's approach. In every stratum, families producing tobacco have less land available to them, per adult equivalent. The community-wide means show non–tobacco growers have three and a half times the land that tobacco growers have, and the difference between these means is statistically significant at the .05 level, using a paired t test. As with Chayanov's ratio, however, the picture presented is quite different if the large landowners are removed from the

TABLE 7.4.   TOBACCO PRODUCERS AND THE MEAN LAND/
            ADULT EQUIVALENT RATIO BY STRATUM

Stratum	Tobacco producers	N	Not tobacco producers	N	Mean for stratum
Landless	.35	5	.39	8	.38
Heirs	.5	6	1.80	2	.89
Small	.87	13	1.17	10	1.00
Medium	2.57	13	3.27	4	2.74
Large	5.71	1	19.77	7	18.02
Whole community	1.47	38	5.48	31	3.27

calculation. The average number of manzanas per adult equivalent then becomes 1.35 for tobacco growers and 1.31 for non–tobacco growers, obviously an insignificant difference.

Boserup's hypothesis is equally unhelpful in predicting who will grow tobacco. Table 7.5 again divides the community into two equal groups, omitting the large landholders. Of the 31 families with less land available per adult equivalent, 18 grow tobacco and 13 do not. Of the 30 households with ratios over the midpoint, 19 plant tobacco and 11 do not. It is important to note that of the 37 tobacco producers, 18 of them have low land/AE ratios, while the other 19 are better off. Even among the poorest families with only one-half a manzana per adult or less (an amount barely sufficient to meet their food needs), only 15 of the 26 households have decided to grow tobacco. Thus, while there is clearly some relationship here between agricultural intensification, land resources, and family needs, the patterns are not simple, based on a single "key" variable.

FLOW CHART OF DECISIONS

By combining both the variables of access to land and household labor resources with an assessment of risk for the poorer households in the community, a more successful prediction of land use choices is possible. Though land resources were not found to be predictive for the tobacco decision in the previous section, they were found to influence strongly the whole range of

TABLE 7.5.   TOBACCO PRODUCERS BY LAND/ADULT EQUIVALENT
            RATIO, NUMBER OF HOUSEHOLDS
            (LARGE LANDHOLDERS OMITTED)

Producers	Ratio .83 and under	Ratio .84 and over	Totals
Tobacco producers	18	19	37
Not tobacco producers	13	11	24
Totals	31	30	61

land use choices discussed in Chapter 6 (Table 6.1). The second major factor, the presence of a man under 60 in the house, was shown in Table 7.3 to be an important variable in the tobacco decision. These two factors do not account for all the aspects of land use patterns in Paso, however, and the intervening variables, such as risk, are discussed later for each stratum. Figure 7.1 presents the crop options that would be predicted for each stratum based on the preceding information. These predictions are shown in solid lines, while those choices that deviate from expectations are shown in dotted lines. The numbers beneath each crop show the number of households planting that option. The number of choices correctly predicted can thus be calculated by the difference between the number of households in the group and the number actually planting the expected crops. (See Figure 7.1.)

Landless farmers would be expected to assure their subsistence needs in corn and beans, and to do so in rotation with tobacco, thereby producing more grains and maximizing their cash income as well. Since tobacco is expensive and risky, however, and since few landless families have access to good land for rental from large landholders, it can be predicted that some will opt for the safer traditional grains production. The assessment of risk is a subjective one and involves each household's experience with tobacco and its land, labor, and capital resources. It is hard, therefore, to make a flat prediction of how many landless families will choose tobacco. It can be expected, though, that any family without a man under 60 would avoid tobacco.

Figure 7.1 upholds these predictions and shows that all rented land in this stratum is used for either tobacco or traditional grains. Landless farmers do not consider coffee or pasture. As shown in Table 5.1, profit per manzana of pasture is too low to meet the family's needs. Coffee, a permanent crop, would also be impossible, given the shifting market of land rentals; the landless farmer could not be sure the land would be his in two years, when the coffee was mature. None of the landless households is lacking a man under 60, yet they split sharply over the tobacco decision: five plant it and eight do not.

What characteristics distinguish the households that decide to plant tobacco from those that opt for traditional corn and beans? Those who plant tobacco have access to slightly more land; they average 1.7 manzanas total, while the non–tobacco growers average .8 manzana. As shown above in Tables 7.1 and 7.4, the tobacco producers also have a higher consumer-worker ratio (4.35 versus 3.14 adult equivalents per male worker) and very slightly more favorable population pressure ratio (.35 versus .39 manzanas per adult equivalent). None of these factors, however, can be used alone to determine the landless household's willingness to enter into tobacco production.

FIGURE 7.1. FLOW CHART OF LAND USE DECISIONS

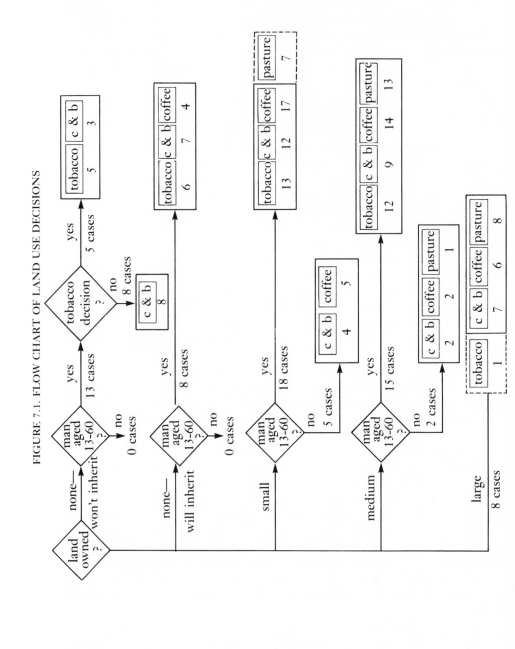

My discussions with landless families revealed the complexity of the tobacco choice for such poor households. Access to appropriate land is one major issue cited by several farmers. For instance, one landless man decided against tobacco because the land offered him for rent had previously been in sugarcane and would have been excessively hard to terrace because the soil was so packed. Other informants felt they would be unable to use the land for more than one year and were unwilling to do the extremely hard labor of a first terrace without reaping the benefits of that labor in succeeding years. The investment in a tobacco shed presented a similar problem if the land rental were not secure. Also, rents for tobacco were high—up to ₡400 per manzana. Since beginners often had low profits, many landless farmers were afraid of the high rents.

Other criteria were important as well in the tobacco decision. One landless man said he decided to grow tobacco because the tobacco cooperative offered loans during the growing season that he could use to support his family, since he would be doing less wage labor while he was growing tobacco. This problem of supplementary income during the growing season was a deterrent to other families who did not have access to such credit or who did not want to risk such large loans. In other cases, landless men had ties to specific large landholders and were reluctant to give up the steady wage work there, for fear of being unable to get it back should they fail with tobacco. Landless farmers weighed their abilities to grow tobacco successfully. One man vacillated because he wanted to improve his marginal standard of living, but he found tobacco very *dificultoso*. For 1973, he decided to stay with traditional grains.

The landless families are closer to economic disaster because they have no cushion to fall back on. Risk-taking behavior is thus more difficult for them than it is for most Pasanos. Lipton points out that the uncertain natural and social environment within which many peasants operate creates a situation in which maximization of efficiency or profit without regard to possible loss is dangerous. Private loss is not buffered by public protection and "thus the risk of harvest failure . . . assumes immense proportions" (Lipton 1968:334; see also Johnson 1971b:149; Cancian 1972:143; Wharton 1971:171). Risk taking may also have long-term effects, since one bad harvest with ensuing lowered nutrition for the whole family may leave effects for many years. Johnson also stresses the importance of security, noting that agricultural and social investments may be made with less than optimum rewards to maximize security (Johnson 1971b:143–144).

One landless informant summed up this environment of decision making.

I would like to plant tobacco because it is a good crop that makes lots of money. But I am all alone [no children]. Since I usually have to leave my rented plot after one year, I would lose all the work of the terracing, and it is such hard work the first year. It's better to leave tobacco alone and stick to corn and beans. Though one can't make much money from them, one can't lose that much either. And one can do it alone.

It is ironic that the landless vary in the tobacco choice. Their economic situation demands the highest possible return per land unit, but this same economic situation makes them least able to sustain the risk and obtain the resources for tobacco. The landless in Paso most need the benefits of this innovation yet are precisely the ones with the most constraints against undertaking it, a situation not uncommon throughout history.

The heirs can be predicted to choose a crop mix similar to that of the landless, but since they know they will be inheriting land some day, they may feel they can make greater investments in the farm. Pasture gives too low a return for these landless farmers, but the solid returns of coffee should be more attractive. Again, households without a man under 60 can be expected to avoid tobacco.

Figure 7.1 shows that of the eight households in the heirs stratum, none lacks the labor resources for tobacco, and all but two plant tobacco. The two farmers who avoid tobacco are brothers, and both cite a reluctance to go into debt and to have dealings with banks as the reasons for their decisions. It should be noted, however, that both families are small, and both men have access to considerable land from their parents. Neither has ever worked with tobacco, nor has their father, a large landholder.

All but one household in this stratum plants traditional corn and beans as well, and half of the families maintain a small stand of coffee. The variability in the decision to plant coffee depends primarily on the inheritance situation of the family. Some heirs know which piece of land they will probably receive when their parents die and so they feel safe investing in coffee. Others live on plots that are not favorably located for a stand of coffee, and thus they prefer to wait to make that investment.

Small landholders would still fall short of the farm size needed for significant investments in pasture, but they can be expected to use all three of the other options. In fact, Figure 7.1 shows that the small landholders fall into two land use types, on the basis of adequate family labor resources. Families with no male under 60 plant traditional corn and beans and coffee. Only one case of the five, a widow who lives entirely off her small stand of coffee, diverges from this pattern. Small landholders with insufficient labor resources are the farmers noted in Chapter 6 who evaluate tobacco as highly profitable, but not the crop for them. Their labor constraints lead them to coffee instead.

The remaining fifteen small farmers generally plant a three-part crop mix of tobacco, traditional grains, and coffee. The variation in this pattern is caused by several factors. First, there are five households who would be predicted to plant tobacco but do not. One is a widow (with grown sons) whose husband planted several manzanas of fruit trees to assure her income after his death. With the security and ease of her fruit production, she does not need to shoulder the risk and hard work of tobacco. Two other households are very small families (two children each), and four of the five who avoid tobacco have access to considerably more land than the rest of the stratum. These various household resources and needs are weighed together, and coffee is preferred as an alternative source of good profits. For all the small farms in this stratum, only one household has chosen to plant no coffee on their land.

Eight of the 18 households have a small plot of pasture as well. These plots, as seen in Table 6.1, average only a little over one manzana and are used in most cases to support a milk cow. Often these bits of pasture were inherited or already sown in purchased land. Only in a few cases was land deliberately put to such a small piece of pasture by its current owner. The families who maintain it, however, obviously value the milk in their diets.

Medium-sized landholders span a range of farm size from 7 to 28 manzanas, sufficient to choose any of the four options available. In fact, they show less variation in their land uses than do small landholders. Two households have no man under 60, and neither plants tobacco, as would be predicted. Pasture becomes common in this stratum—14 of the 17 medium landholders invest land in pasture. All households but one have a stand of coffee. The general pattern is a four-part mix, spreading their land over all the land uses common in Paso. Such a mix spreads peak labor times, risks, market fluctuations, and weather shifts and takes advantage of the strengths of each of the four options.

Large landholders, as discussed in Chapter 6, can be expected not to grow tobacco but to dedicate themselves instead to grains, coffee, and pasture. Since tobacco is the only choice affected by a labor constraint, the presence or absence of adult male workers should not divide this stratum. The flow chart shows that all households but one do, in fact, have the same crop choice pattern. As noted in the last chapter, several large farmers have land in forest and fallow as well, but these uses are not included in the flow chart. All the households have pasture, all but two have coffee (one is a newcomer to the community who has not yet tried to turn a section of pasture over to coffee, and the other just cut down his old coffee because it was unproductive). As noted, one household has just bought land and continues tobacco production to help pay off those debts. The farmer in this case has three grown sons to help him with the work, though, and says that he looks forward to when he has paid off his farm and can give up the hard work of tobacco.

SUMMARY

In spite of these diverse variables that influence household decisions, the crop mix of Pasanos can be predicted surprisingly well from the land and labor constraints on each household. In the flow chart, 69 households have a choice of four crop options, a total of 276 possible land use decisions. The patterns in the figure correctly predict 230, or 83%, of the actual land uses in Paso. The largest single divergence from the predicted crop mix comes from the expectation that families who plant tobacco will also plant an additional field of traditional corn and beans. While a majority in each stratum does conform to this pattern, a sizable minority rely only on the terraced grains. Such a rejection of traditional corn and beans reflects the larger size of their tobacco plots or their smaller families and lower grain requirements. Since in each case they continue to grow corn and beans, the choice is really one of alternative methods toward the same crop goal. By eliminating these 16 cases as incorrect crop mix predictions, the flow chart accurately predicts 246 out of 276, or 89%, of all land use choices.

Although land and labor resources are the major factors influencing farmers' decisions, the 11% of land use choices that diverge from the predicted pattern reflect a wide variety of considerations. Several variables were tested for their ability to reduce this variance—years of marriage, kin ties, neighborhood groupings, access to credit, experience with new technology, the AE/males ratio and the land/AE ratio—but none proved useful to explain more than one or two cases. Pasanos themselves often use personality or character to explain farming differences: "He has bad habits [vices]," or "He's always liked to work that hard." Such interpersonal variations cannot be discounted, nor can personal preferences for certain crops be ignored.

Also important is the fact that land use decisions are constrained by the farm's history. A farmer who buys land with a stand of coffee or a good pasture on it has to take these previous land uses into account in weighing the costs and benefits of each option. The amount of time since land purchase is also important. One Pasano with a big family was found to have planted one of his four manzanas of land in tobacco and left the other three in forest; on the surface, a most inefficient use of his resources. In fact, however, he had just purchased this land, had no sons to help him, and was able in the first year to clear and plant only the first manzana. His land use reflects a natural process of farm development, and it will take him several years to bring the whole area into crop production.

Thus, when making land use decisions, each household in Paso must balance a unique combination of land resources, labor constraints, and other factors to meet household needs. No one aspect of agricultural production has precedence over all others for all cases, though the preeminent importance of land resources was demonstrated in the last chapter and the role of labor resources added in this one.

Further, it has been shown that some patterns of agricultural decisions are not part of the emic understandings of the farmers themselves. The two heirs who avoid tobacco production because of their distrust of banks and credit cannot be distinguished on that criterion from a whole range of Pasanos who share that distrust but who get tobacco loans anyway. Their family size and access to ample land is more useful to the outsider in explaining the heirs' land uses. Some farmers say the demands of a large family are important in the tobacco decision, but this factor was not found to be a decisive criterion either. This point will be discussed more in Chapter 9.

Tobacco is seen as "very profitable" by some Pasanos while others prefer coffee for generating cash. The preceding analysis has shown that the latter group often has no man under 60 in the house. Informants never discussed the lack of adequate labor resources as a factor in their decision; instead, they stated that tobacco was simply too risky or too arduous. As noted previously, this comparison of emics and etics does not prove these farmers "wrong" in citing risk or hard work—tobacco is generally recognized to have both these characteristics. Rather, the weights attached to these aspects of tobacco vary according to household resources. For the purpose of predicting farmer decisions, the existence of certain labor resources provides a clearer line than some attempt to define which households are "risk averters" or which are "hard labor avoiders."

8

Patterns and Implications of Agricultural Credit

In addition to the new crops and crop mixes in Paso, one of the biggest agricultural changes is the use of credit in the production process. It is now common practice to obtain fertilizer and other inputs on credit, to be repaid after the harvest. Pasanos refer to these loans as *la jarana* ("hassle" or "scrape"), saying "*No quiero meterme en esta jarana*" ("I don't want to get mixed up in that mess"). Although many farmers would rather avoid credit, 84% of all households in Paso had some kind of loan in 1973.

This chapter will explore two major questions: what are the patterns of agricultural credit use in the community, and what are the impacts of this change for all sectors? These questions stem in particular from the experiences in many areas of the world in which dependence on agricultural production credit has exacerbated the disadvantages of poorer farmers and has increased risk and financial disaster for those poorer households that were able to obtain the scarce loans (Baker 1973; Franke 1974). High default rates have led in some areas to foreclosures and increased outmigration to the cities (Frankel 1969). For several reasons, these adverse consequences of agricultural credit programs have not been reproduced in Paso.

METHODOLOGY

To obtain data on credit patterns in Paso, I discussed with each family its history of credit use. I was able to compile information on all but 8 of the 75 households in Paso. In the formal part of the interviews, I asked for information on sources of credit, cosigners, interest rates, and other specifics. The interview usually became informal later as we discussed credit in general, the pros and cons of various sources, and the farmer's individual experience. The following discussion, then, draws from a wealth of anecdotes, opinions, polemics, and case histories, as well as from statistics for each household. I also discussed credit with each of the sources that Pasanos use—the bank, National Production Council, fertilizer distributors, and others. These conversations verified that the procedures described to me by the farmers were reported the same way by the merchants and government officials who administered them.

TYPES OF CREDIT

This chapter will deal primarily with credit for the four crop options: corn, coffee, tobacco, and cattle. The first three options require short-term credit, mainly for fertilizer, with repayment due after the harvest. Loans for cattle and tobacco sheds are made for a longer period, usually four or five years. Pasanos also obtain credit for the purchase of land and for building or repairing houses. Consumption credit is quite rare, but a few families also obtain loans from stores or private individuals for consumption purposes. The procedures for obtaining land, housing, and consumption credit are discussed briefly here; their significance will be discussed in Chapter 9 as well.

Pasanos do save money, but no Pasano I knew of used a bank account. Money is kept at home, and since households are vulnerable to theft, savings were not discussed openly. Even wealthy farmers, however, complained of being short of cash much of the year; most of each family's profits are ploughed back into land, cattle investments, or consumption purchases. Savings will not be further discussed here.

Both men and women obtain loans in Paso. Sometimes women take out loans directly for their own use; one such example was an unmarried woman (living with her mother) who bought a cow. Other times, women get loans so that their husbands can cosign for them. There are a number of cases of the reverse as well, in which the wife cosigns for the husband's loan. In general, a very small proportion of credit in Paso goes to women, fitting my earlier generalization that the agricultural enterprise is considered the husband's domain.

Agricultural production presents four kinds of cash needs: money to buy fertilizer and other inputs, to pay peons, to buy livestock, and to live on until the harvest comes in. Since the price of fertilizer or cattle represents a high cash outlay for most families, fertilizer and cattle are usually financed with credit. Borrowing to cover living expenses or to pay peons, however, is quite controversial. Some farmers welcome these funds as a great help, while others criticize them for leaving the agriculturalist with nothing at the end of the harvest. Most Pasanos prefer to undertake as little debt as possible, and therefore finance only their fertilizer.

There are three kinds of lenders. First, there are private companies that sell fertilizer either on credit or for cash. Second, there are several kinds of government programs, most of which are administered through the local bank, also run by the government. The third source of credit is other Pasanos. These sources of credit vary according to the purpose of the credit, and it will be simpler to outline the options open to Pasanos one purpose at a time.

Loans connected with the production of *corn* can be obtained both from government programs and from private fertilizer companies. There were four different government programs available to Pasanos during the year of this study, but two of them were shortly phased out. Without going into great detail about their differences, all four gave credit primarily for fertilizer and required payment at the end of the year. Two programs required the farmer to find someone to cosign for him, while two programs were set up especially to help the poorer households who were likely to have difficulty finding a cosigner. Some lenders require the signature of a cosigner, who must have a certain value of assets, or else the lenders require that the borrower put up his or her own assets as collateral on the loan. Almost all Pasanos use cosigners rather than putting up assets as collateral. Government programs made up roughly 60% of the value of all corn loans in Paso in 1973. In three programs, once the farmers had signed the papers they were given cash to purchase fertilizer from the fertilizer agency of their choice. In the other program, the Ministry of Agriculture made a bulk purchase of fertilizer and distributed it to the participating farmers.

The remaining 40% of loans for corn were made by the private fertilizer companies themselves. There are six such "agencies" in Puriscal, and their prices and brands vary. Each year, Pasanos shop around, seeking the best price for the fertilizer formula they prefer. They show no patterns of loyalty to one agency, nor to one brand of fertilizer, though out of the four or five brands available in town one is avoided and generally believed to be inferior. The process of arranging for credit with the fertilizer companies involves filling out a company order form or a legal paper similar to that used for buying land. Almost all such agreements require a cosigner.

Coffee fertilizer is obtained in a very similar way, with one difference. In Paso, only one company purchases coffee from the producers, and it also finances and provides fertilizer for those who wish. About 15% of the value of coffee loans in 1973 was obtained from this coffee company. The remainder of coffee credit was obtained from the fertilizer companies. For coffee, the choice to use agencies rather than the coffee company has more to do with the conditions of the loans than with the price. When coffee is sold during the harvest season, the coffee company will not pay the full price if the farmer has a debt with them. Rather, a proportion of the sale price is paid and the rest applied toward the debt. By the coffee harvest season, many Pasano households are anxious for cash and prefer a form of credit that leaves them more freedom in paying off the loan.

Until recently, *tobacco* was financed through the tobacco companies. Farmers received vouchers from the tobacco companies, and they took the vouchers to a fertilizer agency to obtain fertilizer and insecticide. During this research, however, a new system was established whereby the farmer's tobacco contract serves as collateral for a regular bank loan. Almost all tobacco farmers arranged for such a loan—some for funds sufficient to cover cash inputs and others for additional cash to pay peons or to build a tobacco shed.

There are a few miscellaneous cases in Paso of loans being obtained to plant beans and one case each of vegetables and rice. This credit was obtained only through government programs and usually involved some cash for labor. In the case of beans, the seeds were also a major expense. Some of these loans required a cosigner but others did not.

Cattle are financed exclusively from banks, but about one-third of these loans were made from a bank in San José, the capital city, while the other two-thirds were from the usual Puriscal bank. The terms and requirements are slightly easier at the bank in San José, and some households also got credit there because they felt they were reaching the credit limit at the local bank. The term of the loan varies with the number of cattle purchased. A new feature of cattle loans allows a grace period of one year before payments begin, and several Pasanos welcomed this change as very helpful. All cattle loans require a cosigner.

Interest

Virtually all loans in Paso, whether from the banks, government programs, or fertilizer agencies, are made at 8% interest for the loan period, regardless of purpose. The interviews show 82% of all loans discussed in the histories were made at 8%. Another 9% of cases recorded show an interest rate of 6%, which was the earlier standard rate in this area. In the remaining 9% of loans discussed, other interest rates were cited, but most of them were cases where the farmers hadn't asked but "guessed it was ———" or where the farmer was simply mistaken (for instance, the bank does not charge 12% interest on a corn fertilizer loan).

Houses are financed only through the bank and constitute a relatively new reason for borrowing. Bank records show house loans becoming more common in the 1970s, replacing a short-lived government program for house loans. Houses financed with bank credit are usually painted, modern Type

III houses (see Chapter 3). In the credit histories obtained for Paso, 11 house loans are currently being paid off, most at 8% interest. All house loans require a cosigner.

Credit to buy *land* has the longest terms of all. Land sales are included here as a form of credit, although Pasanos do not see them that way. When an agreement to sell is made between two parties, the length of time in which the land will be paid off and the annual payment are agreed upon. At that point, the land is considered sold, and its control passes to the new owner. However, since he or she is still making payments on the land—sometimes for as long as ten years—this arrangement can be seen as credit. Almost all land bought and sold in Paso is handled directly between the two parties, but there are six cases in the credit histories where bank loans were obtained in order to buy land. The reasons why a bank loan was obtained are not clear in most cases, but it generally seems that the seller wanted cash in a lump sum and would not wait for payments over several years. Bank loans use the land being purchased as collateral and will foreclose if the loan is not repaid. Fifteen households in Paso are now making payments on land.

Moneylenders and Informal Credit

Gonzales-Vega (1973) reports that moneylenders and their high interest rates were of concern to early rural credit programs in Costa Rica around 1914. "With the passage of time, moneylenders ceased to be much of a 'problem'." (Gonzales-Vega 1973:43). This pattern seems to be true for the Puriscal region. Pasanos mentioned that in the distant past, there had been wealthy people who acted as moneylenders, but data on current loans from private individuals were hard to find.

Although it was stated to me that one of the older, wealthy men of the community "rented money," I could find no one who had borrowed from him. It was clear that, in emergencies, wealthier families who had cash on hand lent it to their families or relatives, sometimes charging interest, sometimes not. One and a half or two percent interest per month was quoted most commonly as "what they charge," but most people interviewed were opposed to such borrowing and felt it was too expensive. I have no evidence that private loans for small sums are common among friends or relatives. However, this was a subject Pasanos did not willingly discuss, and there may be patterns of borrowing I was not aware of. While it would be difficult to say that there is "enough" capital available for farmers' needs, the minor role of moneylenders in the community does support the conclusion that the availability of credit from the bank and other agencies at 8% interest has removed most of the need for informal credit.

The advent of the highway and the regular bus service has changed the patterns of weekly shopping in the last ten years, leaving credit from stores of minor importance in the economic life of the community. Paso's main storekeeper claims that before the highway most household shopping was done at his store, and farmers even purchased such expensive items as tools and saddles there. Many of these transactions were on credit, and some farmers sold their harvest to the same storekeeper, thereby cancelling the debt. Today, this same store is used mainly for emergency purchases, such as when a household runs out of salt. Most shoppers prefer to go to Puriscal, where goods are cheaper and of higher quality and where variety is better.

Only 10% of Pasanos reported that they ever took goods from the general store on credit. Often, these transactions involve nothing more than a quick purchase while the householder has no money on him or her, and the money is paid later in the week or soon thereafter. Six families (9%) stated that they have used credit in stores in Puriscal, but all say these transactions are rare and involve small amounts of money. In general, then, consumer purchases in Paso and in Puriscal are made in cash, and Pasanos seem to prefer it this way. No formal interest rate is charged for consumer credit.

PATTERNS OF CREDIT USE

The credit types just discussed are of unequal importance in Paso, and the patterns of their use reveal the preeminence of loans for tobacco and cattle, the two primary land use changes. Credit patterns will be analyzed here only for the present because reliable data for the past are not available. The bank is unable to keep records of individual credit transactions after they are paid off. Most of the fertilizer agencies are relatively recent establishments and also do not carry back records. Therefore, reconstructing credit patterns in Paso five or ten years ago would require reliance on Pasanos' memories. While I have confidence in their ability to remember how old a coffee stand is or how long ago they began using fertilizer, remembering the exact amount of a loan five years ago, or the interest rate, or even the number of years needed to pay it off, is much more difficult. Therefore, this analysis is constrained to look only at current cases. Consumer credit is not included here because the amounts are relatively insignificant.

Table 8.1 shows the total outstanding loan transactions in 1973 as reported in the credit interviews. In addition, it shows the colon value of payments due for each type of credit in 1973. Thus, the total loans for annual crops are included here, but for items such as tobacco sheds or houses only the annual payment due is included. These figures are an estimation of the total outflow of capital as payments on loans from Paso in 1973. Because of the missing interviews with eight households, the figures are not complete, but they give an adequate idea of the distribution between uses of credit in Paso. (See Table 8.1.)

In terms of numbers of transactions, the annual loans for corn, coffee, and tobacco (91 cases) make up 64% of the total credit arrangements for the

TABLE 8.1. TOTAL CREDIT BY PURPOSE, 1973

Purpose	Number of transactions	Value of total 1973 payments due
Corn	37	₡ 10,464
Coffee	24	7,181
Tobacco	30	29,667[a]
Cattle	29	20,485
Houses	10	8,500
Land	13	51,700
Totals	143	₡127,997

[a]At the time of the credit interviews, many 1973 tobacco contracts had not yet been signed; therefore, where 1973 is unavailable the 1972 figure is used, leaving the tobacco total slightly lower than it would otherwise be.

year. With the 29 (20%) for the purchase of cattle, these loans greatly outnumber the 23 transactions for houses and land. In terms of the value of the payments due in 1973, land is by far the largest use of credit, comprising 40%. As noted, however, land payments are not usually seen as a form of credit by Pasanos. Excluding land payments, the colon value of tobacco and cattle transactions emerge as the major types of credit in Paso. Corn production uses about half as much credit, and coffee still less. The figure for coffee may be slightly low since some households had not yet bought their fertilizer for the second application in 1973. Many of these households, however, planned to skip this second application because the price of fertilizer had gone up so much. I estimate that the total for 1973 would not increase by ₡1,000 over the ₡7,181 shown.

Credit Use by Stratum

Agricultural credit is widely used by all strata in Paso. Loan funds are not monopolized by the wealthiest families as in some developing countries (Franke 1974; Gotsch 1973; Griffin 1974; Mencher 1970, 1978), nor are landless families excluded. Table 8.2 shows the patterns of credit use by stratum. (See Table 8.2.) The table excludes loans for land and houses and presents the data on credit for the four land use options only. More than half of all the five strata used credit in 1973, from 63% of the large landowners to 100% of the heirs. Even though the large landowners presumably have access to considerable capital, they feel it is worth their while to borrow money for these purposes, and they average 2.2 transactions apiece. The average number of transactions per household with credit shows that the landless families average 1⅓ transactions, and most of these loans are for fertilizer. Heirs and small farmers show more, 2.1 and 2.7 transactions respectively, while medium-sized farmers average 2.9 transactions per

TABLE 8.2.   AGRICULTURAL AND LIVESTOCK CREDIT USE
BY STRATUM, 1973

Stratum	Total number of current transactions	Average number of current transactions per household with credit	Percentage of stratum using credit	Number of households interviewed
Landless	14	1.3	79[a]	14
Heirs	17	2.1	100	8
Small	40	2.7	79	19
Medium	38	2.9	93	14
Large	11	2.2	63	8
Totals	120	1.9	77	63

[a]This figure excludes the three economically inactive households and the two households who were unable to obtain land to rent in 1973.

household, the highest in the community. When linked with the fact that 13 of the 14 medium households interviewed used credit, this stratum can be seen to be taking the greatest advantage of these programs.

Table 8.2 illustrates that all strata have access to and use credit; what proportion of the credit money goes to each stratum, however, and how equitably is it distributed? Table 8.3 compares the percentages of the colon value of agricultural and livestock credit by stratum with the percentages of each stratum of the total population and the total land used. Population is measured as adult equivalents, to make household needs more comparable, and land used by each stratum includes both land owned and rented. The colon value of credit for tobacco in 1973 includes the estimates based on 1972 figures, as in Table 8.2.

Table 8.3 shows that landless and small farmers obtain less credit for agriculture and livestock than would be expected from their numbers in the community. On the other hand, the landless, heirs, small, and medium

TABLE 8.3.   COMPARATIVE CREDIT USE BY STRATUM, 1973

	A. Percentage of total Paso population (in AE)	B. Percentage of total value of credit (in colones)	C. B as percentage of A	D. Percentage of total land used by Pasanos	E. B as percentage of D
Landless	20	12	60	2	600 (+10)
Heirs	10	13	130	3	433 (+10)
Small	29	23	79	7	329 (+16)
Medium	28	32	114	22	145 (+10)
Large	14	20	143	66	30 (−46)
	101	100		100	

farmers all obtain more credit than their "share," if based on the amount of land used. Column C shows the first trend and expresses as a percentage the ratio of each stratum's credit share and its proportion of the population. Only the landless and the small farmers receive a lower percentage of credit than is their percent of adult equivalents in the community. This difference, however, never exceeds 8 percentage points, indicating that the imbalances in credit access are not large. When this imbalance is taken together with the fact that landless households average only 1.3 credit transactions while all other strata average more than 2 (as seen in Table 8.2), it seems clear that the landless have somewhat less access to credit. Possibly they have less inclination to obtain credit, but in any case, the heirs, medium households, and large households get a disproportionate share of the loan amounts obtained by the community.

Column E shows the opposite trend, in which all four strata except for the large landholders receive more credit than their percentage of the total land used. In terms of the productive capacity of the land in Paso, these four strata are using more capital and are more involved with credit institutions to achieve their current agricultural production. This latter difference is, of course, the result of the relatively low capital cost of pasture and the relative underutilization of land on the large farms. Nevertheless, the table mirrors a pattern of capital intensification that follows the labor intensity discussed in the previous chapters. Not only are the smaller farmers terracing and triple cropping their lands in tobacco production, but they are also investing more money in order to obtain their higher yields and profits. To the extent that all credit involves risk, these figures also show a greater willingness on the part of the four strata to absorb risk when compared to the large landholders.

IMPLICATIONS OF CREDIT USE

Access to Credit and Social Stratification in Paso

Two aspects of credit may be a deterrent to the poorer farmers in Paso: the risk involved and the necessity of obtaining a cosigner. Many of the points made in Chapter 7 to explain why some landless farmers avoid tobacco production are pertinent here to the risks involved in credit. Where family incomes are already too low and access to both rental land and wage labor opportunities is precarious, there is little buffer against disaster. This situation may raise farmers' assessment of the risks involved in borrowing money and thereby leave them reluctant to join the available programs. This assessment may explain why both landless and small farmers obtain less than a proportionate share of credit, while heirs, who may have access to more land and parental support, obtain more than their share. Nevertheless, the fact that only 21% of these poorer families did not have loan transactions in

1973 shows that the deterrent of risk affects a small proportion of these households.

Turning to the issue of obtaining a cosigner, social stratification does affect credit use in this way. A borrower is required by most lenders to present either collateral or a cosigner. Each lender has different requirements for the cosigner's assets: some require landownership with a deed or other collateral in excess of a certain value. Cosigners take a risk that they may be legally required to pay the debt of the borrowers if for some reason they are unable or refuse to repay.

All strata discussed the process of finding a cosigner as problematic. The request to cosign puts a strain on the relationship between two people, and, given the atomistic structure of the community, the cosigner has little to gain from agreeing to help another person in this way. Every Pasano can discuss cases where the cosigner ended up paying the loan. Four cases were recounted to me of Pasanos or their relatives repaying a dept for which they had cosigned. None of the defaulters was from Paso, and a total of four cases in living memory does not seem to present a high risk. Nevertheless, the risk is present in the minds of both parties when loans are undertaken.

It is difficult to determine if the problems of obtaining a cosigner keep some households from obtaining credit they would otherwise like to have. No Pasano specifically stated that he or she had tried and failed to find a cosigner. Such an admission would be embarrassing, of course, and although my informants freely made other embarrassing revelations, perhaps this topic was more sensitive. Another possibility is that Pasanos avoid asking someone to cosign when there is some doubt about the cosigner's willingness, thereby making refusal rare. A third possibility is that although many people are reluctant to ask someone to cosign for them, in fact, a refusal to cosign is rare. Whichever situation is the case, Pasanos prefer to avoid having to obtain a cosigner. One women was attracted to a government program to finance milk cows precisely because the loans would not require a cosigner. Subsequently, the program was changed and this women was bitter because she then had to find a cosigner. Nevertheless, she did find one and was able to purchase her cow.

Patterns of Cosigning

The requirement of a cosigner and the financial risk that it entails for two families provides an interesting insight into the relations between the strata in Paso and the ways in which families handle these credit ties. Table 8.4 shows the types of cosigners used in all the agricultural and livestock loans discussed in the interviews (both current loans and those paid off) in which the relationship between the debtor and the cosigner was known. Cosigners categorized as "close relatives" are the borrowers' siblings, parents, children, spouse and spouse's parents, and siblings and siblings' spouse. The borrow-

TABLE 8.4.   TYPES OF COSIGNERS BY STRATUM OF BORROWER

Type of cosigner	Landless		Heirs		Small		Medium		Large		Totals
Close relatives	3	(17%)	20	(67%)	29	(46%)	28	(53%)	15	(71%)	95
Distant relatives	1	(6%)	2	(7%)	14	(22%)	12	(23%)	0	(0%)	29
Nonrelatives	6	(33%)	2	(7%)	13	(21%)	7	(13%)	3	(14%)	31
Collateral	0	(0%)	0	(0%)	1	(2%)	3	(6%)	1	(5%)	5
No cosigner needed	8	(44%)	6	(20%)	6	(10%)	3	(6%)	2	(10%)	25
Totals	18(100%)		30(101%)		63(101%)		53(101%)		21(100%)		185

er's children's spouses are also considered "close relatives." "Distant relatives" includes all other kin, for example, cousins or spouse's cousins.
Cosigners that are "not relatives" are neighbors, *compadres*, or friends. Some
loans were obtained through the use of collateral—either land, cattle, or a
house. And for some loans, no cosigner was needed.

There are some clear differences among the strata in Paso in their
patterns of cosigning. The landless households use relatives less than one
might expect (23%). Most frequently, they use loans where no cosigner is
required (44%). Thirty-three percent of their loans are cosigned by nonrelatives, but, surprisingly, no loans are cosigned by patrons. In sum, for the
landless households, loans without a cosigner or cosigned by a nonrelative
make up the majority of loans (77%).

Heirs, fittingly, select cosigners primarily (67%) from among close
relatives, with somewhat heavier dependence on affinal than consanguineal
relatives. This figure supports the idea that heirs are close to their wealthier
relatives, though they may not yet participate in that wealth.

Small landholders use all five ways of obtaining cosigners, paralleling
the diversity of crop mix seen in Chapter 7. Forty-six percent of their
cosigners are close relatives, but distant relatives are also used (22%). As
could be predicted, this group of landholders can use collateral, but there is
only one such case. Medium landholders also show a similar variety in choice
of cosigner, but 76% are relatives. As with the small farmers, they use some
nonrelatives (13%) but relatively few loans with collateral or with no cosigner
needed.

Large landholders mainly use close (mostly consanguineal) relatives—
71% of cosigners. They are the only group to show such great reliance on
close kin. Distant relatives are not used, while 13% of their loans use
collateral.

Several patterns emerge from these data. Landless farmers use credit
without cosigners more than any other group and use relatives only 23% of
the time, compared with 73%, 68%, 76%, and 71% for the other four strata.
These poorer families obviously cannot turn to their families for support in

these risks in the same way the rest of the community can. The use of collateral to back up a loan is avoided by all groups. Nonrelatives are cosigners for the landless in one-third of the cases available, are rarely cosigners for heirs (7%), but are found among the other strata with somewhat greater frequency (21%, 13%, and 14%).

The use of credit involves ties between close relatives in over half the cases shown in Table 8.4. Adding distant relatives brings the total to 67% of all cosigners. For a community with relatively few familistic traditions, relatives are obviously becoming more important as credit is more frequently used. Given that the request to cosign a loan is somewhat burdensome to both parties, using a blood tie possibly lessens the difficulty of the request. It may also lessen the freedom the prospective cosigner may feel to deny the request.

Table 8.5 explores the ties between strata by showing the strata of both borrowers and their cosigners. All cases of agricultural and livestock credit obtained in both 1972 and 1973 for which the strata of both parties are known are shown in this table. In addition, cases in which the cosigner is from the same household (a spouse or child) are separated. (See Table 8.5.)

These figures show the wide variety of strata of cosigners used by all groups of Pasanos. Heirs and landless are asked to be cosigners least frequently, as would be expected since many do not have sufficient assets to qualify in some programs. Small and medium households make up the bulk of cosigners, in proportion to their numbers in the population. Medium and large landholders use household members more than other strata, in 33% and 54% of their credit transactions, respectively.

Three patterns that emerged in Chapters 1 through 3 are reflected here in credit use. The majority of the households spread their requests for cosigners, using relatives, neighbors, and friends from all social strata. This atomistic pattern parallels the fluid market of land rentals and peon-employer relations. At the same time, relatives do form the most important

TABLE 8.5.   BORROWERS AND COSIGNERS BY STRATUM, 1972–1973

Cosigners	Borrowers					
Stratum	Landless	Heirs	Small	Medium	Large	Totals
Landless	2	1	3	2	0	8
Heirs	0	1	0	3	4	8
Small	3	6	17	5	1	32
Medium	2	4	18	16	1	41
Large	0	7	7	1	0	15
Same Household	0	0	2	13	7	22
Totals	7	19	47	40	13	126

group of cosigners, similar to the kin bias noted in access to land. Pairs of men or groups of three sometimes cosign for each other, and I noted seven such groups in Paso; almost all were made up of close kin. These cooperative ties bind together heirs and landholders, but no landless farmers are found in these groups.

The third pattern is the absence of a strong economic role for patrons or landholders who regularly hire peons. Of all the cases in which a large landholder cosigned for someone in another stratum, in only one was the borrower his peon. In all other cases the borrowers were either kin or neighbors. This pattern holds too for medium and small landholders who hire wage labor; they rarely serve as cosigners for their peons.

Thus, landless and small landholders get no advantage when obtaining credit from any peon-employer ties they may have. The landless are also less likely to be able to form a cooperative group, borrowing and cosigning for each other. Together with the risks already noted, these factors illuminate some of the disadvantages felt by poorer landholders with regard to credit. The data also show, however, that non-kin Pasanos are willing to help out these farmers by cosigning their loans, and the government programs that do not require cosigners also ease this situation considerably.

Repayment Problems and Credit Default

As small farmer loan programs have expanded in developing countries, default and foreclosure have become serious problems in some places. One of the purposes of my credit research was to determine if these problems accompanied the increase in the use of credit in Paso. Brown (1972) states that for Costa Rica as a whole repayment problems in the rural credit programs have been rare, and on the average only 15% of the loans have received extensions. Evidence from Paso parallels this national pattern. Of 69 households in the credit interviews, only 20 discussed ever having had problems repaying any of their loans. The vast majority of these cases involved crop losses from weather damage. Half of the cases said they had been able to pay on time, usually by using profit from other crops or by selling what the family would normally eat. Some households had a calf or a pig they sold, and one man sent a daughter to the city as a maid and thereby obtained the cash for repayment.

Eight of the 20 cases with problems were late in repaying. In general, the bank and other programs have been willing to extend the due date on loans when Pasanos are unable to pay on time, and by this means have gained a relatively calm relationship with the farmers of the community. Some of these households who are late now may eventually default—the programs are new and these effects not completely clear. At the time of the research, however, these farmers fully intended to pay.

Two cases of the 20 with credit problems involved default. In one, the household head was sick and could not see how he was going to repay the loan; he had given up. What the bank will do in this case is not yet clear. The other case is the only example of bank foreclosure in the history of Paso. The incident occurred before the current bank programs began, and the bank's conduct in this case is discussed by most Pasanos as completely fair. Considering the large number of loans in recent years and the vagaries of weather, health, and agricultural skill, that there have not been more serious problems with loan repayment is significant.

Pasanos show considerable concern about their credit ratings. If someone should default on a bank loan, he or she is put on "the black list" and cannot obtain further loans. Several very poor farmers have expressed concern over repaying debts and guarding their right to future credit. In only one case found during this research did a man living near Paso choose to default. He borrowed money to plant corn, the harvest went poorly, and he asked for a second year to repay. The next year was an adequate harvest, but he said, "I'd have to sell all of this year's crop to pay the loan and that would leave us with nothing to eat." So he planned to default: "What else can we do?"

The absence of widespread repayment problems does not, however, hide the risks involved, especially for households closer to the subsistence margin. For them, back-up resources to cover weather damage and other problems are not available, and loan repayment directly affects the standard of living. As one farmer who lost his harvest but paid on time said, "Even though we don't eat, we have to pay up."

When in 1973 the bank began a new program that was designed to attract new borrowers, it offered loans without cosigners and with a form of crop insurance: if the harvest failed through no fault of the agriculturalist, the loan would be canceled. This program is the first one to offer crop insurance and is especially attractive for this reason. As this research ended, a large number of Pasanos were planning to enroll in this program, and poorer farmers were particularly anxious to reduce their risk in this way. Were crop insurance regularly available for all crops at a reasonable price, Pasanos would welcome it and pay gladly.

The Role of Credit in Paso

Credit programs such as this one described for Paso have had significant effects, both positive and negative, on the national level in Costa Rica. Many Costa Rican bankers were opposed at one time to expansion of credit to small farmers because they felt credit "had merely been an instrument for maintaining the level of production of small farmers, rather than a tool for change" (Gonzales-Vega 1973:81). While it is true that credit for fertilizer use does generally act to maintain the productivity of lands that can no longer be left to fallow, this maintenance of productivity is no small feat. Given the rapid population growth in the community and the pressure on the lands available to the majority of Pasanos, replacing soil fertility with chemical fertilizers has allowed this population to continue a productive life in agriculture. The fact that credit and tobacco production in particular can at least partially counteract the increasingly unfavorable prices and market structures in the rural areas (as discussed in Chapter 4) has allowed Costa Rica more years than most Latin American countries without facing the political and economic problems of rural-urban migration and swelling slums.

Some economists would argue that the Pasano minifundia are uneconomic and that displacing a portion of the population would increase efficiency of both the plots and the labor force. The Costa Rican situation, however, makes this point of view unrealistic. Industry has yet to become a major employer in Costa Rica, and there is no surplus of urban jobs. People forced off the land have nowhere to go and would most likely end up in a life of urban poverty and marginal employment. If more remunerative uses for their labor appear, Pasanos will move, but as a number of them have discovered, life in the city now provides a lower standard of living than does life in the country. These return migrants prefer life in the rural area, and credit has played a role in maintaining the viability of that choice.

The overall productivity of the community has increased from the use of credit. If fertilizer were only available for cash, it would be out of reach for the majority of landless, heirs, and small farmers. Even medium-sized farmers say they are short of cash much of the year. Through the use of fertilizer, the community produces more tobacco, corn, beans, and coffee than it otherwise would, and much of this production is sold. Credit has also undoubtedly stimulated the purchase of beef cattle by wealthier Pasanos. Since Costa Rica earns scarce foreign exchange with exports of tobacco, coffee, and beef, this support for the productivity of the community helps in the national balance of payments and releases capital for other uses.

If credit programs are successful from the banker's point of view, they result in further earnings for the urban sector from the rural hinterland. With 8% interest rates, the relatively high costs of administering loans to small farmers, and the rate of inflation in Costa Rica, however, these programs may not be financially successful in terms of generating profits for the bank. They may, therefore, take the form of a small subsidy to rural production costs, which may slightly offset the soaring costs of fertilizer and the unfavorable terms of trade for many crops produced.

These national-level benefits of credit have not come without costs, however. Production credit has added a new kind of dependence to Paso's dependence on world markets for the sale of produce. Not only must Pasanos watch beef and tobacco prices and policies in the United States and other countries, but they are also subject now to the pressures of fluctuating oil prices. For the nation as a whole, fertilizer use represents not only a drain on foreign exchange but a new kind of vulnerability. The agricultural production of most of the households in communities like Paso is now dependent on purchases of petroleum-based fertilizer. Just as the farmer in Chapter 1 stated "we are trapped," so too is the country as a whole trapped by the rising costs of this nutrient subsidy to its soils.

Most Pasanos, however, do not discuss the international or even national aspects of credit use. The many conversations I had with farmers and their families reflected the more short-term, concrete considerations of the costs and benefits of credit. Some statements were very positive.

> Credit is a very good thing for me. When I am really hard up, a storekeeper or a banker or someone will lend me something so that I don't want.

> Thanks be to God that there is credit.

> If one can keep up with the payments, it is what agriculturalists here need most and is very good for us.

> It's a great thing. Because the interest rate is better than borrowing from moneylenders.

Some were quite negative.

> I'm an enemy of these hassles; I don't want people pointing me out as the one who won't pay his debts.

> It's very bad to operate with bank loans. The bank people come out and look at your cows or your crops and tell you how to work or criticize or take away your cow. Also, applying for it is a waste of time and no one pays you for it. And the money on the interest is lost.

> Credit can be very bad, especially for people who drink up all the money and end up owing.

> Credit basically leaves one going around in circles—paying up and returning to borrow again.

And some statements were mixed.

> Sometimes it hurts one, but sometimes one can do OK.

> The bank is much better for farmers now and is very helpful. But it's often hard for the landless to get a loan, which makes the poor poorer in the long run.

Credit is basically good, because it helps you out in times when you have no money. It would be better, of course, to have a regular income and have money coming in all the time, but in place of that, credit helps.

Of 50 such discussions about credit, 22 were completely positive, 13 were completely negative, and 15 were mixed. Looking at these attitudes toward credit by stratum, there is some tendency for small and medium-sized landholders to make positive comments about credit and for large landholders to make negative comments. These tendencies fit the credit pattern of large landholders who use credit less frequently than the other landed strata.

One small landholder had a whole philosophy about community development and credit.

Credit doesn't really help the poor. What the government can do if it really wants to help the poor is to buy up a huge farm and divide it to sell, comfortably, to the poor without land. Sell the land at comfortable prices—a manzana perhaps. Not that it has to be good land, but one can work the soil from one side to another [a reference to tobacco terracing] and thus live. A place to build a nice house—that's what the government should do, not give loans.

Take José, for example. He could rent only a piece of bad land for corn—all he could do was burn the low scrub on it. He put in work and fertilizer and got very little harvest out of it—left with his debt to repay. He ended up worse off!

Poor people shouldn't try to use loans for more than fertilizer, because they have nothing left afterwards. Now, the new bank program is a help—it's a good idea. But if you have to sell part of your corn to pay for the fertilizer and if you had a loan to pay labor too, you'll always stay poor. And if you then have to pay rent on top of that—it's all gone. It's hard for a poor person to get stronger. It's better to help the poor to be able to work for themselves.

Pasanos' diverse opinions reflect both the positive and negative impacts of credit programs on the community level. Among the positive effects is the tendency for credit use to counterbalance the ruinous effects of soil depletion and land shortage and thereby help maintain a minimal standard of living for those families most hurt by these trends. The use of credit is one part of the farmer's response to the ecological imperative of population pressure. At the same time, credit use both stems from and results in a greater involvement by institutions and forces from outside the community. Thus, credit involves both of the forces for change outlined in Chapter 1.

Credit has allowed another sector of families to increase their standard of living through investments in tobacco and beef production. The rising consumption standards in dress, diet, and housing detailed in Chapter 3 are supported by these investments. Credit also acts as a "tool for change" in agricultural methods. The relative success of these programs has encouraged a degree of optimism among the larger farmers, and all strata are trying new ideas and taking new risks. Far from just maintaining the status quo, agricultural credit in Paso has supported a new level of capital investment and technological innovation and has clearly benefited a majority of families who have used it.

Why have Pasanos' experiences with credit resulted in relatively few of the disasters noted in other countries? The answer lies in five general characteristics of Paso. First, there has been no "crash" Green Revolution-type program in which all farmers were urged to adopt new methods and credit in one large package. Technological change has come relatively slowly and gradually and has been adopted by each farmer at his own pace. More vulnerable farmers have often hung back, while others absorbed the high risk of early innovation (Cancian 1979; Berry 1980).

Second, early credit programs financed only fertilizer, not labor costs, thereby lowering the total loan amount. Third, banks have, for the most part, treated their customers fairly and have been generous when farmers have been unable to pay on time. This good rapport has helped farmers value their credit ratings and has helped to safeguard the programs from distrust and massive default. Fourth, Pasanos are almost all literate and do not have a history of being deceived or cheated by literate outsiders. Their responses to the credit programs have been based more on the nature of the programs themselves than on past dealings with outsiders.

Finally, and perhaps most important, the agricultural technology and innovations supported by credit are successful. Cattle purchased with bank loans do give a profit to the farmer; likewise tobacco production and fertilizer use all repay the farmer for the expenses of the inputs. Unlike areas where "miracle seeds" did not germinate (Frankel 1969) or new varieties did not respond to local conditions (Winkelmann 1976), agricultural changes in Paso have been successful for most of the farmers who adopted them.

The local-level costs of credit are clear, however. Several families suffered considerable hardships in order to repay loans after harvest failures. The increased risks of credit are particularly threatening to the strata that are least able to withstand them. The new crop insurance program, if continued and extended to all annual credit, could provide an important balance to this increased risk.

The use of credit for cattle can also been seen as a cost to the majority of the community. As we have seen, much of the land shortage that has unfavorably affected many households is due to the expansion of beef production. Adverse ecological impacts, both local and nation-wide, would have advanced more slowly, perhaps with time for more careful assessment of their consequences, had beef production not received infusions of capital.

In addition, the dependency on national and international prices and supplies adds a new level of uncertainty to farming. Pasanos have no control over the costs or availability of their inputs and find themselves watching passively as credit programs come and go, determined by policies in the capital city. But in spite of the drawbacks to the local community, large numbers of households have responded quickly to these new credit opportunities. *La jarana*, therefore, is a mixed proposition—a hassle, a risk, a sticky web, but also *buenísimo*, "one of the biggest helps to the farmer there is."

9

Household Consumption Goals and the Law of Least Effort

The intensification of land use is one of the most important agricultural changes going on not only in Paso but in rural areas all over the world: peasant farmers are responding to scarce land by changing their land uses toward increases in labor, capital, and technology. More harvests per year, more yield per harvest—these are the goals of agricultural intensification.

This same process of change in Paso provides insights into two theoretical perspectives on agricultural development. First, in Boserup's theory, intensification is seen to imply a fall in returns to labor, and therefore is resisted by farmers who follow the "law of least effort." Boserup's theory operates on a macro-cultural level, specifying that population pressure forces traditional agricultural patterns toward intensification, but she does not specify how individual farmers are motivated to overcome the law of least effort. Data from Paso can be used as an example of intensification in process and provide evidence that farmers continue to follow the law of least effort, even when intensifying. The evidence from tobacco production and traditional grains production shows that soil depletion and the lowered labor efficiency of the traditional methods lead toward the higher efficiency of the more intensive methods.

This evolutionary perspective is congruent with Chayanov's theory of peasant economy, which postulates the importance of returns to labor in household decisions. Yet Chapter 7 showed that many medium landholders who have no need to grow tobacco continue this intensive land use. Data from Paso show that a rising standard of living for some households and a desire to increase both consumption standards and the farm's assets push these farmers toward intensification. Tobacco is a means of upward mobility in Paso, and the decisions involved with its production illustrate that not only a household's land and labor resources but also its needs and goals must be taken into account to understand the decision to intensify (as elaborated in Chapter 7).

This chapter will complete the analysis of agricultural change in Paso by assessing the impact of the Paso data for these two theoretical perspectives. In the process, the two forces for change outlined in Chapter 1 will emerge again, each playing its part in structuring not only these changes in the community itself but also the implications of this case for a better analysis of agricultural decision making throughout the world.

AGRICULTURAL INTENSIFICATION IN PASO

The shift from traditional slash-and-burn techniques for the production of corn and beans to the triple cropping of terraced tobacco production illustrates many aspects of the process of agricultural intensification. Although the agricultural changes in Paso are much more rapid than those usually associated with "evolution" and are more linked to international markets and national governmental institutions, this Costa Rican case nevertheless provides a means of measuring the transition process in a way that would be impossible in more conventional situations of intensification.

Intensification is defined by Brookfield and Hart as "a measure of the addition of inputs up to—or beyond—the margin where application of further inputs will not increase total productivity" (1971:90). These inputs may take such forms as more labor for finer cultivation, more organic nutrients added to the soil in composting, more water in irrigation, or more capital in purchased insecticides or fertilizer. Conklin (1961) stresses a definition of intensification that focuses on the length of the fallow period. These two approaches are connected, since a decreasing fallow period usually requires an increase in labor and other inputs for each plot.

As noted earlier, Boserup and Chayanov associate intensification with declining returns to labor:

> A population which has reached the critical level of density . . . may be compelled . . . to accept more steeply diminishing returns to labor . . . and may have to do longer and harder hours of agricultural work in order to avoid a fall in nutritional standards [Boserup 1965: 41–42, 75].

> Professor E. Laur, for example, has investigated Swiss farms with little land. These farmers trebled their intensity. They suffered a big loss in

income per labor unit, but they gained the opportunity to use their labor power fully, even on the small plot, and to sustain their families [Chayanov 1966: 7–8].

In five ways, then, tobacco production in Paso shares the characteristics of agricultural intensification. First, it reflects a sharp decline in the fallow period, from shifting cultivation to triple cropping. Paso's traditional corn and bean production (for those households that can still rotate their land with fallow) fits Pelzer's definition of shifting agriculture (Pelzer 1945): fields are rotated and cultivated less time than they are in fallow; they are cleared by fire; human labor alone is used; there are no draft animals nor is manuring practiced; fields are not plowed; and the primary tools are hoes (machetes) and dibble sticks. By Brookfield and Hart's typology, Paso's traditional production was Class 1A: simple swidden-type cultivation (1971:105–109). The bean production described in Chapter 4 is similar to the "slash-mulch" cultivation in the Pacific lowlands of Colombia (West 1957:129). As has already been described, the methods practiced by most Pasanos today for their traditional corn and beans are a truncated swidden system: fields can no longer be rotated and thus are rarely burned, but the same tools and planting techniques are used. Those who have experienced a sharp drop in soil fertility have begun to apply chemical fertilizers and thus have begun the process of adding new inputs to the old methods. With tobacco production, Pasanos get three harvests per year and have moved into Boserup's category of "multiple cropping."

Second, labor investment has increased as the land is used more intensively. Table 9.1 compares the average labor input in Paso's tobacco and grains rotation to other labor intensive systems for which data are available. The figures are all converted to eight-hour days per hectare. (See Table 9.1.) Table 9.1 shows that the amount of time spent on tobacco and grains in Paso

TABLE 9.1.   COMPARISON OF LABOR INTENSITY: PERSON/DAYS^a PER HECTARE

Location	Crop	Person/days
Paso	Tobacco and grains	332
Paso	Tobacco only	239
China	Tobacco only	217 (Buck 1964:302)
Gambia	Rough rice	192 (Clark and Haswell 1964:90)
Thailand	Paddy rice	131 (Hanks 1972:60)
China	Paddy rice	126 (Buck 1964:302)
India	Paddy rice	125 (Boserup 1965:51)
Paso	Grains after tobacco	92

^{a}Eight hours/day.

is much greater than the time typically spent on wet rice, nearly three times the number of workdays measured by Buck, Hanks, and Boserup. Even for the cultivation of tobacco alone, labor input in Paso is higher than for wet rice, although similar to tobacco production in China. Within Paso, labor investment in tobacco production is four times that of the traditional grains.

Third, as discussed in Chapter 8, tobacco production involves significant increases in capital inputs, especially for fertilizer.

Fourth, the technology invested in tobacco is reminiscent of the changes associated with intensification. Although tobacco terraces bear only a small physical resemblance to the permanent stone terraces of many wet rice areas (not to mention the stone terraces of the Incas), nevertheless the mountainside is carved up into stair-step ridges of earth far more extensive than the "quasi-terraces" in Peru described by Brush (1977:98). The soil quality is changed by the annual remaking of these dry terraces and the composting of the tobacco and corn stalks, and the labor investment involved is enormous. In addition, permanent tobacco sheds must be built and sprayers must be purchased. The complex procedures of transplanting tobacco seedlings, fertilizer and insecticide application, pruning, and harvesting may not rival the intricacies of water management in rice terraces, but they do involve a big change from the methods needed for traditional grains.

Finally, the returns to labor in tobacco are sharply below those of the less intensive grains production. Table 9.2 presents again some of the data from Table 5.3 in Chapter 5. Thus, as both Boserup and Chayanov would predict, a day's labor in the traditional land use rewards the worker with roughly 40% more profit than either of the parts of terraced agriculture.

THE MECHANISM OF AGRICULTURAL EVOLUTION

Given that tobacco production in Paso bears considerable similarity to agricultural intensification as defined by Brookfield and Hart, Conklin, Boserup, and Chayanov, we can now explore what motivates farmers to overcome the law of least effort to adopt these new methods. As discussed in Chapter 5, families are pushed into tobacco from land scarcity and from a desire to maximize both cash return and grains on small plots. These reasons

TABLE 9.2. RETURNS TO LABOR OF TOBACCO AND
TRADITIONAL GRAINS (IN COLONES PER MANZANA)

Crop	Return
Traditional grains	₡17.25
Tobacco and grains	₡12.15
Tobacco only	₡11.62
Grains after tobacco only	₡13.54

suggest that farmers simply settle for "more effort" because household needs are not met by traditional means. In fact, the process is more complex.

Harris, following Leslie White (1959), has looked at the issue of the evolution of agriculture in terms of energy use and energy efficiency (Harris 1971:205). Energy as a unit of measure allows us to combine the labor input into a specific process, with the harvest resulting from that labor. Since the focus of Boserup's theory is on labor investment, labor efficiency in energy terms is a useful measure for comparing traditional and intensive land uses in Paso and avoids the distortions of profit comparisons, which reflect market prices.

Harris uses the formula:

$$E = m \times t \times r \times e$$

to compute the caloric efficiency of a particular land use system. E stands for the number of calories a system produces annually; m is the number of food producers; t is the number of hours of work required; r equals the calories expended per hour; and e is the average number of calories produced per calorie expended. For our purposes, we need to find e for both systems in Paso. To do so, it is unnecessary to compute the *total* number of calories that traditional or intensive land uses produce (E). Instead, we can compare the total number of calories produced *per manzana* per year. Since Chapter 5 presents data on jornals/manzana, m and t can be combined. Harris uses a standard figure of 150 calories per hour for the intensity of labor (r). Thus, the labor efficiency of each system (e) can be calculated by dividing the total number of calories produced per manzana per year by the total number of calories expended per manzana per year to produce it.

Beginning with traditional grains, this calculation is straightforward. Table 5.6 shows that traditional corn yields are 73 cajuelas per manzana. Multiplying this quantity by 33 pounds per cajuela times 1,641 calories per pound of corn—Leung (1961) presents these figures for Latin America—gives a total production of 3,953,169 calories of corn per manzana. The same calculation for beans is 6.5 quintals per manzana times 100 pounds per quintal times 1,532 calories per pound, or 995,800 calories of beans per manzana. The labor invested in corn and beans averages 72 jornals per manzana, which can be multiplied by 6 hours per jornal and 150 calories per hour to give 64,800 calories expended per manzana. By dividing the total calories produced by calories expended, e for traditional corn and beans production is 76.

The comparison with the tobacco rotation presents two problems. The first is that tobacco is produced for nicotine, not food value, and hence is hard to assess in caloric terms. Therefore, this analysis will compare only the terraced grain production, since these yields are an important reason Pasanos have switched to tobacco. The second problem is that these grains benefit from the tobacco terrace, but it would not be appropriate to apply all the labor cost of the entire year to the grain yields. Since the grains contribute about one-third to the cash value of the year's harvest, it can be arbitrarily decided that one-third of the labor investment in the terrace should be added to the direct labor of the corn and bean production process. Using the same caloric equivalents as in the first calculation, e for terraced grains is 74.

These two figures (76 and 74) are nearly identical and suggest that in terms of labor efficiency there is no advantage to either method. On the basis of these figures alone, we would have to conclude that desire for the increased profit per land unit in tobacco production is what overcomes the law of least effort. The comparison of traditional grains with terraced grains makes one important error, however; it lumps the production of all five strata together. Traditional grains production figures are, as noted, high, since many farmers with lowered yields by these methods have already switched their fields to the terraced production. Only some of the large landholders and some of the landless households who rent from them have sufficient land to be able to rotate their crop land with fallow; these households make up roughly 25% of the farmers planting traditional corn and beans. Although the current yields of this group cannot automatically be assumed to represent the historical fertility of soils under slash-and-burn techniques in the past, they are the closest approximation available today for measurement.

Tobacco terraces likewise improve the fertility of the soil, and therefore the top 25% of these farmers were also separated out, based on the length of time their land has been terraced. The energy efficiency calculations from these four groups—the top 25% of both terraced and traditional grains, and the remaining 75% of each—are shown in Table 9.3. The table shows clearly that for the large landholders and those who can rent from them, the traditional methods of grain production are more efficient (101 versus either 77 or 71). This result supports the evidence presented in Tables 9.1 and 9.2 that less intensive systems give higher returns to labor. In a situation of population pressure, however, where soils have become overworked and depleted, extensive systems do not produce as well. To the remaining 75% of farmers, traditional methods return only 63 calories for each calorie invested rather than the 101 returned to the top 25%. With terraced grains, the well-established terraces produce slightly better than newer ones (77 calories versus 71), but the difference is minor. The important comparison for

TABLE 9.3.   CALORIES PRODUCED PER CALORIE EXPENDED
             IN GRAIN PRODUCTION

Production type	Top 25%	Remaining 75%
Traditional grains	101	63
Terraced grains	77	71

understanding agricultural evolution is that given the poorer soils of the remaining 75%, more intensive methods produce 71 calories per unit of labor while extensive methods produce 63. Thus, the extra labor that terracing requires not only repays the household that suffers low yields from traditional grains but also gives a somewhat higher return than the extensive system.

The Pasano situation thus suggests that declining soil fertility may be important to individual farmers in motivating the shift to more intensive methods. When the decline in fallow period lowers soil fertility and crop yields, the efficiency of labor invested in such a system drops too. In this case, for the majority of the community, the more intensive terraced grains are actually *more* efficient in terms of calories produced per calorie invested. Pasanos who adopt the more intensive methods can therefore be said still to be following the law of least effort.

The sharp differentiation in access to land in Paso creates a situation in which both the old agricultural adaptation and the new can be measured at once. Large farmers continue to follow the high profits and returns to labor of the traditional grains. The effect of land shortage on the majority of the community creates a situation of population pressure and declining yields, changing the decision-making environment. This analysis shows that not only does tobacco production become desirable because of its higher returns to land, it also responds to the declining returns to labor in traditional fields. Here, as noted in the credit discussion in Chapter 8, the greater costs of tobacco production serve to reverse the decline in labor efficiency in the older, extensive methods. In sum, though this energy efficiency comparison suffers from the difficulties of comparability of tobacco and grains and can only be a partial analysis, it supports the theorists who find population pressure a key variable in agricultural intensification, and it shows that in situations of change, farmers may continue to follow the law of least effort to a new level of agricultural evolution.

TOBACCO FOR SUBSISTENCE OR INCREASED CONSUMPTION?

Thus far, we have discussed the adoption of tobacco as a decision motivated primarily by necessity. Tobacco's high costs in labor, capital, expertise, and risk present a considerable deterrent to a casual decision to give it a try, without pressure from growing family needs, scarce land, and declining grain yields. Nevertheless, the flow chart in Chapter 7 shows that a large majority of medium-sized farmers plant tobacco, although they have considerable amounts of land. The question arises here for Paso, as it does in many world areas: to what extent is this agricultural intensification a necessity and to what extent does it act to increase surplus production over basic subsistence needs?

The generation of increased surplus can come about by coercion, through heavier taxation or forced labor in irrigation agriculture (Geertz 1963; Sanders and Price 1968; B. White 1973), or by a spontaneous desire for an increasing standard of living (Rawski 1972). Bennett has pointed out the need to address the effects of "rapidly changing purpose and want factors" (Bennett 1976:850). These new "want factors" play an important role in the tobacco decisions of some Pasano households.

To distinguish farmers who plant tobacco from necessity from those who plant it to increase household consumption, I divided the 38 tobacco growers in Paso into two groups. My data on the cost of the average Pasano diet show that half a manzana can provide one adult with sufficient corn and cash to meet food needs for roughly a year. Expenses for clothing, housing, and medical costs are additional to this half manzana, and it can therefore be arbitrarily estimated that one manzana per adult equivalent is sufficient land to meet basic needs. Families with at least this much land can be said to have no objective necessity to adopt tobacco.

The cutoff point of one manzana per adult equivalent neatly divides the tobacco growers in half. Those families who average less than one manzana/AE are Group I and can be assumed to choose tobacco to meet unmet subsistence needs. Group II, made up of households averaging more than one manzana/AE, is the focus of our interest: why do they grow tobacco?

There are a number of interesting differences between these groups. First, Group II families are a little older than Group I; they average 21 years of marriage, while Group I averages 15 years. In addition, Group II families are quite large. The average household size in Paso is 6.6 persons and 5.1 adult equivalents, but Group II households average 8.0 persons and 6.3 adult equivalents (Group I averages 7.6 persons and 5.7 AE). Many Group II farmers say they like tobacco production because it allows them to use their children's labor. A large family can be expected to reduce the cost of hired workers, and the increased profits for these families may account for their continued tobacco production. The data, however, do not support this hypothesis. The relationship between family size and tobacco profits is low (measured here by the phi coefficient since the sample is too small for the contingency coefficient):

$$\phi = .21 \text{ (not significant at the 0.1 level)}$$

Thus, the value of children as workers cannot by itself explain the tobacco decisions in Group II.

These two groups also differ in the amount of experience they have had with tobacco. Group II averages 9.6 years of working with the crop; Group I has a mean of 6.5 years. The complexity of tobacco technology shows that experience is essential to good tobacco production, and the profits made by

Groups I and II bear this out. Households that have planted tobacco continuously for 10 years or more make an average of

₡3,381 per manzana.

Households with 5 or fewer years of experience average

₡1,650 per manzana.

Thus, the Group II farmers do benefit from higher profits.

Although family size alone does not affect profits significantly, the combination of a large family and longer experience yields higher profits. By separating out those families who are "large" (with nine members or more) and those who also have "long experience" (eight years or more) from both Groups I and II, three distinct groups emerge. Group I households with smaller farms and less experience average only ₡1,450 profit in tobacco. Group II households with smaller farms and less experience average ₡2,415, while the experienced farmers with large families in both groups average ₡3,687. These figures suggest that Group II families, who tend to be more experienced and have more children, continue tobacco production to take advantage of their higher profits.

The histories of the 19 Group II families are also helpful in understanding their land use decisions. By reconstructing their family and farm size when they began tobacco production, it becomes clear that 7 of the 19 households began tobacco many years ago, when they had access to little or no land and needed tobacco's greater profits for subsistence. These 7 households used to be in the situation where Group I families now find themselves; they needed tobacco's high productivity to meet household needs. With the tobacco profits, they were able to buy more land and today all are medium landholders.

Another 9 of the 19 Group II households owned some land at the time they began tobacco but could not have made sufficient profits from that base to be able to buy more. The men of these households turned to tobacco to gain the means to expand beyond their former farm size. Several of these households are young couples whose land resources are adequate now only because their families are still small. They have undertaken to increase farm size in anticipation of future needs.

The remaining 3 cases are all older men who are no longer expanding their landholdings. Each man has at least one grown son who has taken an active interest in tobacco, and therefore these households have no labor limitations. Their profits are used either to support the sons' marriage plans or for other investments on the farm. Thus, 7 of the 19 Group II families have already used tobacco to pull themselves out of the subsistence pressure characterizing Group I. Another 9 are in the process of expanding their farm size, while the 3 older men are acting to increase their incomes beyond their current comfortable levels.

The way tobacco profits are used illustrates these differences between Group I and Group II. Table 9.4 shows the number of households currently making payments for land, cattle, or new houses. Eighteen of the 19 Group II families are in debt for either land, cattle, or house purchases, but only 9 of the Group I families have commitments of this sort. Most Group I families cannot undertake "improvement" expenses as Group II can because they are at present barely meeting their subsistence needs. Thus, Group II families can be seen as continuing their highly profitable production of tobacco in order to pay off the lands they have bought or to invest in cattle or a nicer house.

TABLE 9.4.   INVESTMENTS OF TOBACCO PRODUCERS,
             NUMBER OF HOUSEHOLDS

Investments	Group I	Group II
Making land payments	3	13
Making house or cattle payments	7	15
No outstanding debts	10	1
	19	19

These findings challenge two theoretical perspectives. One, Boserup's law of least effort, would expect the 19 Group II families to resist the extra work of tobacco and use their lands for pasture, traditional grains, or coffee. In fact, this is exactly the goal of many of these farmers. Once their lands are purchased and their pastures stocked with cows, they will have assured themselves a generous land base from which to be able to avoid the labor of tobacco terracing. "I want to put these terraces into pasture soon," said one Group II farmer. "I don't need to buy any more land, and besides—I'm getting too old for all this tobacco stuff." Not only have these farmers escaped a subsistence need for tobacco, but tobacco production will have allowed them upward mobility and a higher standard of living as well.

Chayanov's labor-consumer balance also predicts that Group II families would give up tobacco to lower their degree of self-exploitation, once sufficient lands had been purchased. Chayanov assumes that the standard of living goals of the family will remain constant: "Another less important, yet essential social factor, is the traditional standard of living, laid down by custom and habit, which determines the extent of consumption" (Chayanov 1966:12).

Many Group II families, however, do not want to remain at the "customary" standard of living. They want houses with glass windows and tile floors, vinyl furniture, and television sets. They want to have cattle and maybe a horse or two, and perhaps to educate a son or daughter in the city. This desire for upward mobility, whether in Paso or in the capital city, attacks the foundations of Chayanov's theory of peasant economy. The family's consumption goals must be added to their decision-making equation, and these goals are influenced by the greater contact Pasanos have with the outside and with the new institutions and consumer influences that come into the community.

These same "development" forces that have brought the new highway, the export market for beef, and the ecological and economic dislocations of pasture have also brought the tobacco companies, the credit programs to finance tobacco, and later the upward mobility from tobacco and pasture for a segment of the community. The desire for new consumer goods and a higher standard of living is a part of this development process as well, but the new land uses of tobacco and pasture serve some Pasanos as a means for achieving these new desires.

Thus, the community must be seen as made up of diverse kinds of farmers, each responding to different household needs and goals with different resources. Those who feel population pressure most keenly respond with intensification and show us that soil depletion, lowered yields, and declining returns to labor are important elements in the mechanism of agricultural evolution. The wealthiest farmers can continue their traditional methods of production and have expanded into pasture in order to maximize the overall returns to the farm. These decisions, as we have seen, increase the population pressure on the landless and small farmers. Finally, a group of middle-size farmers forces us to acknowledge the role of changing consumer demand in encouraging intensification of land use and illustrates that this process may in turn allow a degree of upward mobility.

Conclusion

In this study of agricultural change in Paso, I have explored processes of agrarian transformation that affect many developing countries and many rural communities. Before I pursue the implications of such processes for economic development programs, a review of the conclusions of previous chapters is in order. With an understanding of ongoing community processes and the ways they can be studied, I will then turn to the implications of the Paso case for general understandings of both economic development and programs to guide and influence that development.

AGRARIAN TRANSFORMATION AND FARMERS' DECISION MAKING

Agricultural, Economic, and Ecological Changes in Paso

Until the last 20 years, the traditional agriculture in Paso produced corn, beans, and rice with slash-and-burn techniques. A high population growth rate throughout the community's 120-year history has led to severe population pressure. Coupled with very unequal distribution of the land, this population pressure now means that most farmers do not have sufficient land to rotate their fields with fallow. As a result, soil fertility has sharply declined, and most fields require the application of chemical fertilizers to produce adequate yields.

Into this situation have come important changes in transportation, communication, and marketing. A new, all-weather road was built ten years prior to this research, allowing contact with a wide range of services and facilities in the market town and in the capital city. Each day, buses and trucks pass through the community, connecting Pasanos with institutions such as the agricultural extension service, the bank, and the hospital. These new forms of transportation allow the sale of new crops to international markets for coffee, tobacco, and beef and make possible the purchases of new consumption goods as well. The interaction of the internal processes of population growth and the external impact of increased contact with the outside have led to the restructuring of agricultural choices in the community. Pasanos have turned away from their former exclusive reliance on traditional corn and bean production, rotated with forest and fallow, and have now become deeply involved in beef and tobacco production.

The growing market for imported beef in the United States and soaring world prices have prompted vast increases in pasture by large landholders. Partially stimulated by new credit facilities, new cross-breeds of cattle, and new pasture grasses, cattle production has brought greatly increased wealth to a small segment of the community. Pasture provides low returns per land unit, but for farmers with large extensions of land it becomes a profitable land use, given its low risk and low labor costs.

Beef production has profoundly affected the natural environment of Costa Rica. Ecologists and government experts have expressed concern over the massive deforestation in progress. In several areas, including Paso, dry season water supplies are declining, soil erosion has become critical, and farmers claim that even weather patterns are changing. The new pasture grasses are tenacious, and some Pasanos worry that land sown to them cannot easily be returned to agriculture.

Together with the new market for beef exports, however, has come a very different new land use option—tobacco. Tobacco production is complicated and high risk, requiring expensive fertilization, spraying, and storage. Tobacco must be planted in dry terraces, carved by hand from Paso's slopes. A plot of tobacco requires four times more labor per land unit than any other crop. In return it provides high profits and triple the cash return per land unit of competing land uses. Tobacco is doubly adaptive in that the cash crop itself is rotated each year with corn and beans, thereby providing good harvests for household consumption and sale.

Tobacco is ecologically different from pasture. Tobacco terraces are made with the corn and tobacco stalks of previous harvests, thereby composting some organic matter. Together with the heavy chemical fertilization tobacco requires, these terraces produce higher yields of corn and beans than do regular plots in the community. Terracing actually improves the fertility of the soil, and the contour ridges act to resist erosion as well.

Tobacco thus maximizes the return to small plots and has provided a solution to the population pressure and land scarcity in the community. Families require time to master the complex technology of tobacco, but when they do so, their profits are even higher. Tobacco production has allowed one group of families to buy land and therefore escape the squeeze of land shortage. It is ironic that tobacco, whose nicotine is of dubious value to humankind, has been the economic salvation of many of the families in the community.

If it were not for the situation of scarce land and declining soil fertility in Paso, it is unlikely that tobacco would have been adopted as a land use to the degree that it has. Most farmers need this ecological imperative to overcome tobacco's high costs, though some farmers, probably those who are younger and more eager to take advantage of its high profits, might grow tobacco anyway. Tobacco would not be a viable option without the road, however. Not only is it difficult to ship to market by ox-cart but it is cumbersome to get fertilizer in large quantities into the community. The company experts whose supervision is required for the tobacco contract would also undoubtedly be reluctant to journey into Paso on horseback.

Beef production, in contrast to tobacco, is a change in Paso that exacerbates the problem of land scarcity. Lands that were formerly in fallow or forest were available for rental to the landless or small farmers. With the lands of more large landholders now in grasses, these families have to turn elsewhere to meet household subsistence needs. Tobacco and pasture thus represent two very divergent alternatives, both of which bring higher profits than traditional options to some Pasanos. On the one hand, tobacco provides a capital- and labor-intensive option for agricultural development. Pasture, in contrast, uses both land and labor extensively. Since growing populations and increasing contact with national and international markets are characteristic of virtually all developing countries, the implications of these two alternatives are relevant to rural communities elsewhere.

The Implications for Social Stratification

Recent agricultural changes in Paso have contributed to a greater disparity in the lives of the rich and the poor, and a number of aspects of the improved standard of living in the community have increased that gap. Everyone is delighted with the new road and the bus service now possible. But in the past, landless farmers and their wealthier neighbors might walk together to the market town, whereas today some of the poor cannot afford the bus while a few of the rich own private cars. At a time when no one wore shoes, the peon's lower income was less noticeable than it is today, as clothing, housing, and diet have all increased markedly in cost for the majority of the community.

Upward mobility for most Pasanos has become blocked by the scarcity of land. When land rentals were freely available to the landless families at comfortable rents, poorer households could reasonably anticipate that hard work would enable them to save enough to buy land. Rentals are now smaller, harder to find, and more expensive. Since the land rented has become less fertile, often producing lower yields, the landless are increasingly unable to realize sufficient profits to buy land. Their growing helplessness in improving their economic situation is sharpened as they see their own thatched huts become surrounded by the painted, many-roomed, glass-windowed new houses owned by others.

Land scarcity strengthens the domains in which kinship plays a special role. Families who cannot afford to buy land must now depend on inheritances from parents, and it has been shown that landless farmers who will someday inherit land behave differently from those who will not. Kin ties are also used for obtaining cosigners for loans and land to rent, and the obligations felt by many landholders to help kin first can only strengthen the inequality of access to land. In addition, the infrastructural improvements in Pasano life, such as the road and the new electric line, have boosted the value of land, thereby further pushing prices out of the reach of the less well-off families. Land prices over the last ten years have gone up as much as 1,000%. Since land is the basis from which all Pasanos earn a living, this new rigidity in land ownership pushes the community toward a more fixed stratification system.

The growing income disparities in the community have led to a greater polarization between the strata. The right of large landholders to use their land as they choose is challenged by some farmers, who argue that property ownership should not allow a small group of Pasanos to deprive a larger group of land to work. The increase in pasture acreage thus may be a part of the decline in the local political influence of the large landholders. Leadership positions in the community used to be exclusively in the hands of these wealthy farmers. Today, political power and prestige are possessed by several small and medium landholders who act as brokers with outside governmental agencies. These men are all skilled in the complex technology of tobacco production and are therefore respected for their economic abilities as well as for their work on behalf of the community. Although unequal land ownership has been characteristic of Paso since the early days of its existence, the decline in political power and community esteem of the large landholders has created a new distance between the strata.

In many areas of life, Pasano families continue to operate in independent and egalitarian ways, but this atomism is being undermined by the effects of land scarcity. One family that was unable to find land to rent attached itself to a large landholder in a classic patron-client relation, and thereby obtained a plot to work. It is unlikely that a large class of such dependent landless laborers will emerge, however, simply because cattle production on these farms creates few labor demands. It was shown in Chapter 6 that large farmers prefer to dedicate only a small portion of their land to crops, which would provide more jobs for the landless, because pasture is more profitable on the large farm and has much lower risk. If land in the area becomes primarily dedicated to pasture, the region can expect to see the logical results of such a decline in wage-earning opportunities: massive out-migration and a remaining population characterized by a few owners of large ranches and their dependent peons. Guanacaste, Costa Rica's northwest province, has seen this process unfold over the last twenty years.

At the same time, egalitarian and atomistic patterns remain strong in Paso. Many recent land sales and land rentals were made to non-kin, and ties between employers and peons generally remain fluid. Purchases both in Paso and in town involve no traditional loyalty to one store or seller, nor are requests for cosigning made to the same people every year. Thus, economic ties of these types are still not locked into a rigid system of stratification in Paso.

Although wealth and consumption patterns divide the community sharply, they do not negate the general isolation of many households and the willingness of many Pasanos to work hard on community projects. Although familism may be growing in importance, it has not challenged a tradition of cooperation in a range of voluntary activities, which are increasing in numbers. However, these community projects often involve the sponsorship of outside organizations and bring Pasanos into closer contact with people who devalue rural life and who heighten farmers' awareness of their lower status in the larger Costa Rican society. As yet, interactions between the strata do not reflect the deference and distance due to outsiders. Intermarriage among landed and landless families continues to be common, and visiting patterns among the strata are fluid. Nevertheless, the increased contact with outsiders will strengthen the awareness of stratification within the community and will continue to present a challenge to Paso's traditionally egalitarian ethos. Paso can thus be seen as an example of emerging stratification in which many egalitarian, atomistic, and cooperative patterns remain strong.

How Do Farmers Make Agricultural Choices?

The four land use options—tobacco and grains, traditional grains, coffee, and pasture—are weighed by each farmer, but households with different resources evaluate the options differently. Although every farmer's situation is unique, the analysis in Chapter 6 showed that groups, defined by access to land, make distinct land use decisions. Five strata are distinguished in Paso: landless, heirs, and small, medium, and large landholders. All farmers agree on the criteria by which the options are evaluated—profits, risks, labor and capital requirements, and yields—but different strata use these criteria in different ways. Risk, for example, means quite different things to a landless family with few resources to help them through a harvest failure. Large landholders, on the other hand, recognize that tobacco is the most profitable land use but still prefer other crop choices that require less labor. Similarly, the evaluation of the desirability of traditional corn and beans production depends on the history of the farm and its soil fertility and length of fallow.

It was shown in Chapter 6 that when land resources are no longer a constraint, large landholders can choose to maximize the overall returns to the farm. Even for these families, however, the decision is based not simply on one key variable. The issues of risk, labor requirements, and returns to labor can all be seen to militate toward the choice of cattle production for large farmers. Smaller farmers who choose tobacco must forego maximizing returns to labor; the high labor costs of the terraced production cannot compete in cash terms with the other crops. Farmers recognize this and prefer to use a Chayanovian profit calculation that subtracts only cash costs from the proceeds of the farm. Pasanos' own cost-benefit calculations thus are quite different from economists' traditional methods, which impute a wage to unpaid family labor (see Barlett 1980b). Any attempt to simulate their choices or predict behavior in this kind of household economy will therefore run the risk of distortion and inaccuracy if a different methodology is used.

Since tobacco represents such a sharp divergence from traditional farming methods, the reasons farmers undertake it are particularly important. Chapter 7 explored two theoretical perspectives for predicting the choice of tobacco among Pasano housholds. The Chayanovian consumer-worker balance was shown to vary with the tobacco decision but cannot predict it. Family labor resources are important, however, since no household without a man between 13 and 60 grows tobacco. Boserup's theory that land scarcity motivates agricultural intensification can be applied to the terraced tobacco production and also successfully distinguishes between groups of tobacco growers and non–tobacco growers. Like labor resources, however, access to land cannot alone predict which households will choose tobacco. When household access to land is combined with family labor resources (in the form of a man under 60), the resulting flow chart of land use choices in the community successfully predicts 83% of all land uses. The major source of variation from the predicted crop mixes is the failure of some tobacco producers to supplement their terraced corn and beans with an addition of traditional corn and beans. If these choices are not counted as errors, then the flow chart accurately predicts 89% of land use choices.

Land use decisions in Paso, though primarily dependent on land and labor resources, still have areas of greater complexity. For instance, the decision of landless farmers whether to undertake tobacco is not linked to verifiable characteristics of the family, but rather is a subjective process involving the following factors: the assessment of the complex technology, experience with the crop, the risks and difficulties of credit, obligations to do wage labor for others, access to appropriate land, and high rents. Personality and personal history factors may also be involved. This example illustrates how etic measures such as profitability must be balanced with emic considerations of a farmer's own skills and circumstances. Given tobacco's higher profits and higher grain yields, it would be easy to say that all landless farmers "ought" to grow tobacco. Rather than a normative evaluation of their production decisions, this analysis has sought to determine *why* they do what they do. In the process, the analysis has illuminated a range of deterrents that affect this disadvantaged group most strongly.

Two general conclusions emerge that apply to development efforts in other rural communities. Access to land and household labor resources are the primary determinants of land use decisions, and these two variables break the farmers of Paso into groups with similar agricultural patterns and similar standards of living. But the natural desire of social scientists to delineate one "key factor," one "miracle seed" of agricultural processes (Cancian 1977), must take account of the fact that while both variables were significantly correlated with agricultural decisions, neither variable alone was able to predict those same choices successfully. The reality of human behavior is more complex, and a wide range of other variables must be included in order to be able to understand all the decisions farmers make.

By analyzing the agricultural decisions of Pasanos in light of evolutionary theory, changes in this one community can be connected to the long-term sweep of agricultural history. Boserup predicts that farmers will follow the "law of least effort" when presented with opportunities to intensify their land use and will only adopt more onerous methods if driven to do so from population pressure. The notion of population pressure can be used in a situation of private property and land concentration by taking into account individual household resources. Not all Pasanos feel the pressure of scarce land; many large farmers continue the traditional extensive grain production and the even more extensive cattle production. On the other hand, small farmers and the landless feel the pressure of growing household needs and increasingly scarce land and are motivated to intensify their land use. As Boserup and Chayanov predict, these farmers accept lower returns to their labor in order to meet household needs. A calculation of the energy efficiency of the traditional and terraced corn and bean production methods shows that the land use history of the plot and its soil quality are important considerations. In terms of calories produced per calorie expended, traditional corn and beans grown in fields that have not been rotated with fallow are less efficient than the more intensive terraced corn and beans. The wealthy farmers whose land can be rotated with fallow are wise not to choose tobacco, both on the grounds of energy efficiency and on the grounds already noted. But for the majority of the community that feels the population pressure, lowered soil fertility and low yields leave energy efficiency in the traditional methods even lower than with the terraced corn and beans. Thus, the Paso case suggests that where population growth on fixed land resources results in soil depletion, farmers may actually continue to follow the "law of least effort" when they intensify.

There is a sector of the community whose decision to produce tobacco supports theorists who see intensification as sometimes the result of changing wants and consumption goals. The middle-sized farmers who plant tobacco even though they have sufficient land to earn an adequate living without it have used their tobacco profits to purchase land, stock the land with cattle, and build new and larger homes. These upwardly mobile Pasanos have moved themselves out of the group that is holding its own with tobacco production and barely meeting subsistence needs. Their profitable use of the more intensive methods has allowed them to move into a more secure and comfortable economic position.

As land prices rise and rentals become more scarce, it is not clear whether the current landless farmers will be able to use tobacco for upward mobility as other farmers have done. Many of these landless households have only recently started tobacco and hence have generally lower profits. In addition, tobacco prices are not keeping pace with the rapid rise in land prices. Access to less land, poorer soils, and lower yields all affect family incomes adversely. It was also shown that with fewer household resources to buffer this risk, most landless families cannot rely on kin, cooperative credit ties, or even the customary economic support of a patron. On the other hand, new government extension programs and access to credit for fertilizer and tobacco expenses act to enhance the possibilities of this group. The growing numbers of these families and their increasing poverty make it particularly important to determine whether profitable, labor-intensive options such as tobacco production can continue to succeed in overcoming the barriers to upward mobility among the landless.

These points concerning agricultural evolution and the law of least effort show that farmers make decisions primarily on the basis of short-term resources and opportunities. When the returns to labor drop too low in one agricultural system, another option becomes feasible, as seen in the switch to tobacco. There is no evidence that Pasanos base their decisions on long-term considerations of tobacco prices, the availability of contracts, or the security of the market. Given that farmers have little control over such prices and markets but have to feed their families on a daily basis, such long-term considerations are probably inappropriate. Large landholders choose pasture from a similar calculation of returns to the farm, without concern for the long-term ecological consequences to the land or the social consequences to the community. Thus, dependence on imports of petroleum for fertilizer and the dangers of erosion and soil depletion are not yet part of the consciousness of Pasanos as they choose their land uses. Pasanos seem open to listening to such long-term perspectives, but until these processes impact their short-term constraints, there is little evidence they can be simply persuaded to respond to them. The emic view is short and concrete, and many more general issues of national development are too abstract to affect the decision making of farmers in Paso.

Measurement and Methodology

Before turning to the general implications of the Paso data for development issues, it is important to note how the contribution of this study's methodology clarifies the analysis of agricultural change. Although, on occasion, different measures would have been appropriate had later conclusions been foreseen, a number of the procedures used had unexpected payoffs. It was important that I attempt to discuss key issues with all farmers rather than with a subsample or group of primary informants. Several times, my early sense of Pasano reality, based on five to ten discussions, would later be proven false. Not only were my partial perceptions corrected by a more thorough survey of the community, but the complete data allowed the kinds of intracommunity comparisons that reveal the patterns of credit use, land use, and so forth.

In order to check my impressions and provide subsequently verifiable measures for other researchers, I endeavored to gather data systematically and to quantify certain patterns. In several cases, these measures had unexpected results or made more detailed analyses possible. For instance, I originally gathered genealogical data on all households from a desire to understand kin linkages over time. Reconstruction of these charts for all families, however, made possible the longitudinal analysis of population growth, discussed in Chapter 1. In order to quantify Pasanos' contact with the outside, I attempted to develop a measure of frequency of contact with the market town and the capital city, also presented in Chapter 1. In the process, however, some interesting patterns of sex roles appeared, and I gained a sharper sense of the division of labor by sex in Paso.

By using an actor-oriented framework of analysis, I did not expect Pasanos always to share common choices or attitudes. Records of diverse statements about such things as crop decisions, new credit programs, politics, birth control, and education allowed a reconstruction of the community's diversity and helped me avoid a false sense of homogeneity. This effort is part of a wider trend in anthropology that seeks to discern both the forces that affect individual behavior and the ways in which they, in turn, influence group-level patterns that may be labeled "cultural adaptations." Such efforts are only possible where careful measurements expect heterogeneity on the individual level.

Another important aspect of the methodology used here was the distinction between emic and etic perspectives. In much of the analysis, my etic view as an outside researcher was shared by Pasanos themselves, but there were several places in which the approaches were not congruent. For example, some households say, "Beef is the most profitable," while profit figures clearly show that pasture boasts the lowest returns per land unit. This discrepancy is important and shows that large landholders are able to use a different frame of reference from that used by the rest of the community, in which overall returns to their farms are paramount. Since any attempts to reverse the deforestation process in Costa Rica must face the logic of these large farmers, the emic reasons for their high evaluation of pasture are significant.

Another discrepancy between emic and etic criteria occurs when some farmers say they choose to grow tobacco because it allows them to utilize the labor of their families. Etic data show that all sizes of families grow tobacco, and that, overall, the existence of "free" children's labor does not determine this choice. While the consideration of family labor resources is undoubtedly a factor in these decisions, it cannot be shown to be a decisive factor. In contrast, it was shown that households without a strong man to do the arduous work of tobacco do not choose this option. No Pasano, however, mentioned that labor resource as a part of the tobacco decision. Instead, households without such a man simply say, "Tobacco is too much work." Certainly, by all Pasanos' standards, tobacco *is* too much work; yet some households plant it anyway. Thus, the etic measure clarifies a necessary condition for that behavior pattern, a condition that is not part of the emic awareness of people in the community.

One advantage of using etic measures to supplement emic perceptions is the ability to avoid psychological characterizations of actors. Such psychological descriptions imply that behaviors are deeply rooted in the personal makeup and, especially when the topic is rural poverty and growing polarization, give a strong sense of the improbability of change. Such a tendency errs in two directions: it overestimates the tenacity of certain behavior choices, running the risk that the analysis may blame the victim, and it also underestimates the constraints facing the actor and thereby channels energy away from addressing those constraints.

In Paso, for example, one family may say tobacco involves too much risk, while another says it is too much work. Uncritical acceptance of these emic statements may lead to the conclusions that some Pasanos are afraid of risk and others of hard work. As shown in Chapter 7, it can be more useful to look at the resources available to the household. These resources often explain why some famiiles who hate banks get loans anyway while others behave as "risk averters." In the case of two landless brothers, both had access to ample rented land and both had small families to provide for. These men were able to be risk averters with respect to tobacco, while other households were not so fortunate. The same approach clarifies which farmers avoid the hard work of tobacco—they tend to be older men with no grown sons at home.

IMPLICATIONS FOR ECONOMIC DEVELOPMENT
PROCESSES AND PROGRAMS

In discussing the lessons presented by Paso for "economic development," it must first be noted what kind of development we are talking about, with what benefits for whom. As we have seen, the change processes commonly labeled "development" combine complex interactions, with both positive and negative aspects for different groups of people in Paso. Most governments, however, see economic development goals as including a higher standard of living for rural people and a more equal distribution of income. Some experts contend that supplies of nonrenewable resources are inadequate to allow all the world's people an increasing level of consumption. Others stress the unequal distribution of those resources at present. Declining food supplies per capita provide another caution to a naive assumption that a strategy of "growth with equity" provides an easy solution to development problems.

Yet it is precisely because global populations continue to grow while per capita food intake declines, because wealth is increasingly concentrated in the industrialized countries while the Third World falls ever further behind, and because the interdependence of the economic, political, and social life of rich and poor nations grows more intense each year, that the implications of this Costa Rican analysis are important. If we are to increase the world's food production and improve the relative position of farmers in small communities, we must identify both the processes at work in rural areas and the ways in which these processes can be influenced. Regardless of whether interventions are carried out by national government programs, international aid agencies, farmers' cooperatives, revolutionary central planners, or transnational agribusinesses, the lines along which positive changes in the lives of Pasanos are possible must be illuminated. While massive national and international forces constrain the goals of growth with equity, these ideals come closest to expressing the aspirations of Pasanos and, as such, will guide my review of the development implications of this research.

Many aspects of "development" in Paso, such as the highway, the electric line, and new housing, have benefited the landowning majority of the community. Each of these changes, however, brings new opportunities for landless farmers to fall further behind their landed neighbors. Since household income in Paso is derived almost exclusively from the use of land, the one-third of the population without access to this resource faces very different life choices from those with access to land. Such a division within the community is characteristic of farming areas throughout the developing world.

What Can be Done to Help the Landless?

Several possible responses may ameliorate the difficult situation of the landless in Paso. Out-migration is one solution, but the economic prospects for out-migrant families are not bright. Costa Rica's industries are unlikely to be able to absorb these workers, and the urban slums that have appeared in other Latin American countries may be inevitable in Costa Rica as well. One large landholder proposed an alternative solution: "What we need here is a capitalist, someone to bring in an investment of money, to build a factory right here and provide jobs for all these people." Paso's rugged terrain and isolation, however, make it an undesirable factory location. The landless people in the community would prefer a third option—simply to be able to buy this large farmer's land.

The degree of land concentration in Paso makes land reform a viable possibility, though far-reaching political and economic changes would be necessary for such a program to take place. Ample land is available to create a comfortable standard of living for all families now living in the community. In fact, if three-quarters of the land owned by the largest landholder alone were redistributed equally to the landless households, they would all become small landholders with sufficient land to eliminate malnutrition and produce cash crops from a solid economic base. The large landholder himself would be reduced only to a medium-sized farmer. Since there are eight large landholders, land reform could go beyond a simple, one-generation solution. In addition, given the credit programs already in place and the experience many landless families have with tobacco production and new agricultural technology, land reform in an area such as Paso would have a much greater chance of success than in countries without these institutions.

If such a permanent redistribution of land resources remains unlikely, what can be done to improve the situation of the landless? Any action to increase their access to land resources by rental provides a second-best alternative. Pasanos have been shown to be willing to rent much more land than is now available. Their land use choices are affected by the fees charged, and lower rental costs clearly help the household income of the poorest in the community. Since land is often rented from small farmers, though, some legal ceiling or forced reduction of rents may result only in the withdrawal by those owners of their lands from use by others. Tax incentives to large landholders who rent to the landless may be one alternative, while the expropriation and subsequent rental of absentee-owned estates is another.

Renters, especially when they have no security in their access to land, have little incentive to safeguard the long-term fertility of the soil. As seen in developed countries as well, inflation and rising costs of inputs can force renters of land to work the land for its maximum yield, without investments to protect or conserve its soil quality. In addition, the landless in Paso are constrained from trying certain crop options that require more than one year to mature. All these factors suggest that actions to institutionalize longer-term rental contracts will help the landless farmers and protect land resources. For this suggestion also, however, across-the-board legal mandates may have the effect of removing some lands from the rental pool, a trend that obviously must be counteracted in order for the landless families to reap any benefits from legislation concerning rentals.

The landless clearly benefit from higher wages for agricultural labor. In a more polarized situation where they work primarily for wealthy farmers, minimum wage legislation, if enforceable, can be effective in raising household incomes of landless workers. In Paso, where many small and medium farmers—and on occasion even the landless—hire workers, such efforts to boost wages may not result in higher incomes; if farmers cannot afford to increase labor costs, they may simply hire fewer days of help.

Throughout this study I noted that some aspects of the changing agricultural technology acted either favorably or unfavorably on the economic situations of the landless. The new insurance program accompanying bank credit is one example of an effort to make borrowed capital more accessible to the poorest families. There are no easy ways to predict such efforts without careful scrutiny of the ongoing processes in specific rural areas and the potential interactions of technological change. If positive effects on the situations of landless families are desired, improved access to land, credit, technology, and markets all provide opportunities for intervention.

Small Farmers, Increased Production, and Credit

Since this study is not a comparative one, it is impossible to conclude which organization of agricultural production can be the most efficient, productive, or equitable. The Paso data do suggest, however, a number of important characteristics of mixed-crop private farms, under certain economic conditions. These characteristics may be helpful, then, in assessing the most desirable form for agricultural production units.

Given that the world's most productive lands are already in use, increases in agricultural output will now come from intensified uses of that land. The behavior of farmers in Paso provides an important example of agriculturalists' willingness to intensify their land use within a very short period of time. Since land shortage is a problem that affects not only areas of Costa Rica but all over the developing world, this willingness of Pasanos to increase labor investments, even to quadruple the days of work per land unit, is an important consideration for other land-scarce and capital-scarce countries.

Several economic conditions are crucial in allowing this rate of labor intensification. First, the tobacco option was well known in Paso. Its suitability to the climate and soil was established, and the technology needed for terraces, fertilization, and insect control had already been developed. In addition, contracts were given even to producers of very small plots, and credit was also made available in convenient procedures and at fair rates. Farmers felt the pinch of land shortage and growing families and acted accordingly. Their rapid increase in yearly productivity per land unit and the resulting increase in foodgrain security and higher incomes for tobacco producers are testimony to the responsiveness of independent farm producers. As shown by the different patterns of land use by the large landholders, Paso conforms to the generalizations put forward by other researchers that small farmers are more productive and efficient than large ones in their use of scarce national resources (Lipton 1977; Loomis 1976; Rochin 1977).

The intensification of land use in Paso is dramatic when compared with 20 years ago, but it still falls short of truly intensive agricultural systems. The dry terraces for tobacco could be replaced with permanent structures, with attendant benefits for all crops grown. Better water control mechanisms to channel the heavy tropical rains could be installed. Farmers could develop techniques of manuring, composting, and crop rotation that might further protect soil fertility while raising crop yields. Use of more efficient domesticated animals such as pigs or ducks might increase protein resources, while range-fed cattle production methods could be replaced by feed lots and fodder crops. Examples abound, from the Kofyar of Nigeria to the communes of the Peoples Republic of China, from which more intensive land use techniques might be adapted to Paso's ecological conditions. Movements in such directions take advantage of available labor and population pressure while conserving scarce capital and land resources.

The farmers of Paso show several other characteristics relevant to the design of agricultural development programs. They have shown themselves willing to increase their level of "self-exploitation," in Chayanov's terms. Many Pasano families put in longer and harder hours of work today than they did ten years ago, since they see clear benefits from doing so. Increased experimentation and innovation are also noted among these farmers. Since the production methods of each household are distinct, farmers benefit from their neighbors' trials with new corn seeds, new coffee or pasture varieties, and different planting methods. Again, the availability of fertilizer supplies, on credit, and the existence of insecticides, herbicides, and other technological changes that work effectively in the Pasano system facilitate this response to innovation. Farmers were not urged by crash government programs to adopt such changes indiscriminately, as in some Green Revolution areas; they tried these new methods piecemeal and adopted them as they proved useful.

Attention to the productivity of small private plots leads some farmers to tailor techniques to the slope, terrain, or soil of individual parcels of land. This fine-tuning of production techniques adjusts to microclimatic variations and thereby increases yields. Finally, farmers in Paso have been shown to prefer a complex crop mix. This ecological diversity spreads household labor demands and risk as well as providing a more stable economic unit, both to the family and to the nation. All these characteristics of farming in Paso—intensification of land use, increased productivity, increased labor investment and experimentation, a crop mix, and methods fine-tuned to the ecological conditions of each plot—are important factors in any attempt to increase world food supplies.

An important aspect of recent technological change in Paso has been agricultural credit. Eighty-four percent of Pasanos used credit in some form in the year of this research, and their record of repayment and default is unusually good for such programs. Credit for fertilizer and other uses has helped to reverse the trend of declining productivity of this region and has improved the economic chances of many farmers. While credit and the technology it finances present new opportunities for inequality, the vast majority of Pasanos have access to these credit programs and use them to their advantage.

The success of credit programs in increasing household incomes and agricultural productivity can be attributed to their own structure and to the conditions within Paso. Access to credit for landless and small landholders has not been made difficult, and Pasanos' dealings with banks and other credit sources have led to feelings of trust on both sides. While some farmers have faced hardships in repaying loans after poor harvests, the rarity of foreclosure or credit blacklisting contributes to these good relations. The new crop insurance option will help buffer the risks of borrowing, especially for the poorest families. The high level of literacy of Pasanos, the general success of the technology used with borrowed capital, and the ecological pressure of declining soil fertility and land scarcity have all combined to make the record of credit use in Paso an unusual one. It can even be concluded that credit for tobacco production to the landless acts as a counterweight to the inequalities in access to land in the community.

While the use of credit has many positive aspects, it also has liabilities. Harvest failures can now jeopardize the family's welfare for several years, since debts remain to be paid with the succeeding harvest. The increase in cattle production, with all its negative effects for most Pasanos, has been financed to a great extent by credit. Farmers also have no control over these credit programs, and their new dependence on loan-financed fertilizer leaves them more powerless than before in the face of changing governmental policies and priorities.

International Dependence in Pasano Agriculture

Pasanos' dependence on credit combines with their growing dependence on the outside world in a range of other ways. Changes in agriculture have made them into consumers of petroleum, and the sharp increases in fertilizer prices in 1973 were a jolt to their budgets. Some kinds of fertilizer went up in price by one-third, while prices for agricultural products went up little, if at all. With fertilizer the main cash cost for most farmers, this change brought home powerfully the implications of dependence on foreign inputs for agricultural production.

Foreign markets for cash crops are another area of vulnerability for Pasanos' current adaptations. Exports of beef also depend on foreign markets, specifically on the continued suspension of the quota on foreign beef imports in the United States. In the case of tobacco, this dependence is complex, because the cash crop is linked to subsistence production of corn and beans. Like all such cash crops, whether for domestic or foreign markets, market fluctuations depend on forces over which Pasanos have no control. Tobacco may, in the end, turn out to be an ephemeral solution to Pasanos' dilemma. Other tobacco-growing regions in Costa Rica have been devastated by disease, and production has been greatly decreased there. Although most of Paso's tobacco is now used for domestic consumption, exports to the United States have become important in providing new contracts in the community. Costa Rican tobacco is said to be unable to compete with Cuban tobacco, however, and if the United States lifts its economic blockade of Cuba, the export market for Pasano cigar tobacco could well disappear.

In any agricultural production for sale to big companies, some dependency theorists would stress the profits being made by nonproducers, and the consequent exploitation of the rural population. Pasanos would both agree and disagree with this perspective. They are acutely aware of the low returns they receive from tobacco—averaging, as we have seen, less than a dollar a day. But they are also quick to sign up for tobacco contracts since tobacco, its assured credit, and the good harvests of corn and beans afterward provide them benefits they can find in no other crop option or land use. As we have also seen, the greater profits of tobacco have led to markedly increased prosperity for many families in the community. Tobacco thus provides a model for a new cash crop that is uniquely adapted to the economic and ecological needs of a land-scarce rural community. Programs designed to increase foodgrain productivity, soil fertility, and household income would be hard-pressed to match its advantages, despite its clear and well-documented disadvantages.

The new links between farmers in rural Costa Rica and the world markets they now serve have different impacts on each sector of the community. As we have seen, capital intensification and the use of imported fertilizers creates new opportunities for the comparative disadvantage of some households. Exports such as beef, which have beneficial effects for the balance of payments of the nation as a whole, may have unfavorable repercussions on the local standard of living of some sectors. In the case of cattle production in particular, the choice of one group of farmers acts to constrain the land available to the remaining farmers. Ecological concerns raised with pasture parallel the issues of the long-term adaptability of many new agricultural options: as one new choice becomes possible, are many others being excluded as a result? In the case of pasture in Costa Rica, erosion may be excluding a wide range of future agricultural options. The lessons of the Pasano case are clear: different crops, with different technologies and requirements, will have diverse impacts on the income, nutrition, and relative social position of each sector within a farming community. These impacts must be explored before new crop options are designed to conform to development goals.

Paso provides evidence as well of the kind of agricultural conditions that contribute to egalitarian and cooperative ties between households. Stratification and polarization of the community are increased by the extent to which some families are excluded from access to land resources and the extent to which upward mobility is blocked. Despite vast differences in wealth and landownership, the fluidity of economic status in Paso traditionally fostered egalitarian ties between households. Economic cooperation, on the other hand, is more rare than in many peasant farming communities. The facts that all agricultural tasks can be carried out by individual families and that there is no need for community-level coordination of any kind in production matters contribute to a pattern of atomistic relations in the community. Pasanos are accustomed to cooperative efforts on community welfare projects, but cooperation in the sphere of the household economy is rare. This tradition limits their familiarity with forms of social control necessary to effectively organize such efforts as collective farming or the formation of a farmers' union. Even cooperation for group credit or purchases of fertilizer involves new behavior patterns and would require adjustments in accustomed interhousehold ties. Thus, the kinds of agricultural programs devised to help raise the incomes of farmers, especially of landless farmers, will also affect the character and ethos of the community.

Typical of so many villages throughout the developing world, the farm families in this rural community of Costa Rica have been faced with dramatic forces for changes both from the outside and from within. Changing land uses have been shown to affect each sector of the community differently and to bring with them new vulnerabilities to global prices, markets, and government programs. How do Pasanos respond to these new conditions and risks? Households respond in concrete annual decisions, weighing their land and labor resources, evaluating the characteristics of the land uses open to them, and assessing their overall production strategies. Economic alternatives change, as do the amenities and strains of rural life. With increasing contact with the world outside the community and different consumption patterns and values, new family aspirations and expectations are created. These forces will continue to interact with the internal processes of population growth and its ecological effects to persistently restructure the environment of agricultural decisions. It is through the careful analysis of these processes and their effects that efforts can be designed to increase the welfare and relative position of Pasanos and other agriculturalists in rural communities throughout the world.

Bibliography

Acheson, James M.
    1980   Agricultural Business Choices in a Mexican Village. *In* Agricultural Decision Making. Peggy F. Barlett, ed. Pp.241–264. New York: Academic Press.
Adams, Richard N.
    1970   Crucifixion by Power. Austin, TX: University of Texas Press.
Amjad, R.
    1972   A Critique of the Green Revolution in West Pakistan. Pakistan Economic and Social Review 10 (1):17–41.
Arensberg, Conrad M.
    1954   The Community Study Method. American Journal of Sociology 60:109–124.
Azam, K. M.
    1972   The Future of Green Revolution in West Pakistan: A Choice of Strategy. Economic Journal 5 (1):37–59.
Baker, C. B.
    1973   Role of Credit in the Economic Development of Small Farm Agriculture. *In* Small Farmer Credit Analytical Papers. A.I.D. Spring Review of Small Farmer Credit 19 (SR119):41–70. Washington, DC: Agency for International Development.
Banfield, E. C.
    1958   The Moral Basis of a Backward Society. Glencoe, IL: The Free Press.
Barlett, Peggy F.
    1980a   The Anthropological Approach to Fertility Decision Making. *In* Demographic Behavior: Interdisciplinary Perspectives. Thomas K. Burch, ed.

Pps.163–184. Washington, DC: American Association for the Advancement of Science.
1980b Cost Benefit Analysis: A Test of Alternative Methodologies. *In* Agricultural Decision Making. Peggy F. Barlett, ed. Pp.137–160. New York: Academic Press.
Barth, Fredrik
1967 On the Study of Social Change. American Anthropologist 69:661–669.
Basehart, H. W.
1973 Cultivation Intensity, Settlement Patterns, and Homestead Farms Among the Matengo of Tanzania. Ethnology 12:57–74.
Beals, Alan R.
1974 Village Life in South India: Cultural Design and Environmental Variation. Arlington Heights, IL: AHM Publishers.
Bennett, John W.
1968 Reciprocal Economic Exchanges Among North American Agricultural Operators. Southwestern Journal of Anthropology 24:276–309.
1976 Anticipation, Adaptation, and the Concept of Culture. Science 192:847–853.
Berry, Sara S.
1980 Decision Making and Policy Making in Rural Development. *In* Agricultural Decision Making. Peggy F. Barlett, ed. Pp.321–336. New York: Academic Press.
Biesanz, John, and Mavis Biesanz
1944 Costa Rican Life. New York: Columbia University Press.
Blutstein, Howard I., et al.
1970 Area Handbook for Costa Rica. Washington, DC: Foreign Area Studies, American University.
Boserup, Ester
1965 The Conditions of Agricultural Growth. Chicago: Aldine.
Brandes, Stanley H.
1973 Social Structure and Interpersonal Relations in Navanogal, Spain. American Anthropologist 75(3):750–764.
Brookfield, H. C., and D. Hart
1971 Melanesia: A Geographical Interpretation of an Island World. London: Methuen.
Brown, Albert L.
1972 The Agricultural Credit Project of the Agricultural Sector Program of Costa Rica. Country Program Study for AID Spring Review. Vol. 2:1–50. Washington, DC: Agency for International Development.
Brush, Stephen B.
1977 Mountain, Field, and Family: The Economy and Human Ecology of an Andean Valley. Philadelphia: University of Pennsylvania Press.
Buck, J. Lossing
1964 Land Utilization in China. New York: Paragon Book Reprint Corporation (original 1937).
Burling, Robbins
1962 Maximization Theories and the Study of Economic Anthropology. American Anthropologist 64:802–821.
Busey, James L.
1967 Notes on Costa Rican Democracy. Boulder: University of Colorado, Series in Political Science No. 2.

Cancian, Frank
  1965   Economics and Prestige in a Maya Community. Stanford, CA: Stanford
          University Press.
  1972   Change and Uncertainty in a Peasant Economy. Stanford, CA: Stanford
          University Press.
  1977   Can Anthropology Help Agricultural Development? Culture and Agri-
          culture No. 2:1–8. Bulletin of the Anthropological Study Group on
          Agrarian Systems.
  1979   The Innovator's Situation: Upper Middle Class Conservatism in Agri-
          cultural Communities. Stanford, CA: Stanford University Press.
  1980   Risk and Uncertainty in Agricultural Decision Making. In Agricultural
          Decision Making. Peggy F. Barlett, ed. Pp.161–176. New York:
          Academic Press.

Carneiro, Robert
  1961   Slash-and-Burn Cultivation among the Kuikuru and Its Implications for
          Cultural Development in the Amazon Basin. In The Evolution of Horti-
          cultural Systems in Native South America, Causes and Consequences.
          Johannes Wilbert, ed. Pp.47–67. Caracas: Sociedad de Ciencias Natur-
          ales La Salle.

Chapman, Anne M.
  1958   The Tropical Forest Tribes on the Southern Frontier of Mesoamerica.
          Ph.D. dissertation (anthropology), Columbia University.

Chayanov, A. V.
  1966   The Theory of Peasant Economy. D. Thorner, R. E. F. Smith, B.
          Kerblay, eds. Homewood, IL: Irwin (original 1925).

Chibnik, Michael
  1974   Economic Strategies of Small Farmers in Stann Creek District, British
          Honduras. Ph.D. dissertation (anthropology), Columbia University.
  1978   The Value of Subsistence Production. Journal of Anthropological Re-
          search 34(4):551–576.
  1981   Small Farmer Risk Aversion: Peasant Reality or Policymakers' Rationali-
          zation? Culture and Agriculture 10:1–7. Bulletin of the Anthropological
          Study Group on Agrarian Systems.

Clark, Colin
  1967   Population Growth and Land Use. New York: St. Martin's Press.

Clark, Colin, and Margaret Haswell
  1964   The Economics of Subsistence Agriculture. New York: St. Martin's
          Press.

Clarke, William C.
  1966   From Extensive to Intensive Shifting Cultivation: A Succession from
          New Guinea. Ethnology 5:347–359.

Collier, George A.
  1975   Fields of the Tzotzil: The Ecological Bases of Tradition in Highland
          Chiapas. Austin, TX: University of Texas Press.

Conklin, Harold
  1957   Hanunoo Agriculture. Rome: Food and Agriculture Organization.
  1961   The Study of Shifting Cultivation. Current Anthropology 2:27–61.

Cook, Scott
  1973   Production, Ecology, and Economic Anthropology: Notes Toward an
          Integrated Frame of Reference. Social Science Information 12(1):25–52.

Costa Rica, National Planning Office of
  1967   Compendio de Cifras Básicas de Costa Rica: V. Cuentas Nacionales, July.
Davis, Kingsley, and Judith Blake
  1956   Social Structure and Fertility: An Analytical Framework. Economic Development and Cultural Change 4:211–235.
de Janvry, Alain
  1975   The Political Economy of Rural Development in Latin America: An Interpretation. American Journal of Agricultural Economics 57(3):490–499.
  1976   Reply 58(3):590–591.
Denton, Charles F.
  1971   Patterns of Costa Rican Politics. Boston: Allyn and Bacon.
DeWalt, Billie R.
  1975   Inequalities in Wealth, Adoption of Technology, and Production in a Mexican Ejido. American Ethnologist 2(1):149–168.
  1979a Modernization in a Mexican Ejido. New York: Cambridge University Press.
  1979b Alternative Adaptive Strategies in a Mexican Ejido: A New Perspective on Modernization and Development. Human Organization 38(2):134–143.
Diaz, May N.
  1966   Tonalá. Berkeley: University of California Press.
Dirección General de Estadística y Censos
  1970   Anuario Estadística Costa Rica 1970. San José: Ministerio de Economía, Industria, y Comercio.
  1972   Cost of Living Indexes. San José, Costa Rica.
Dumond, D. E.
  1965   Population Growth and Cultural Change. Southwestern Journal of Anthropology 21:302–324.
Durham, William H.
  1979   Scarcity and Survival in Central America: Ecological Origins of the Soccer War. Stanford, CA: Stanford University Press.
Edwards, W., and A. Tversky, eds.
  1967   Decision Making. New York: Penguin.
Epstein, T. Scarlett
  1962   Economic Development and Social Change in South India. Manchester, England: Manchester University Press.
Falcon, Walter P.
  1970   The Green Revolution: Generations of Problems. Cambridge, MA: Development Advisory Service, Center for International Affairs, Harvard. Economic Development Report 154.
Finkler, Kaja
  1979   Applying Econometric Techniques to Economic Anthropology. American Ethnologist 6(4):675–681.
  1980   Agrarian Reform and Economic Development: When is a Landlord a Client and a Sharecropper his Patron? In Agricultural Decision Making. Peggy F. Barlett, ed. Pp.265–288. New York: Academic Press.
Food and Agriculture Organization of the United Nations
  1953   Maize and Maize Diets. Nutrition Division. Rome: Food and Agriculture Organization.

Forman, Shepard
    1970   The Raft Fishermen: Tradition and Change in the Brazilian Peasant
            Economy. Bloomington, IN: University of Indiana Press.
Foster, George M.
    1967   Tzintzuntzan. Boston: Little, Brown.
Frank, André Gunder
    1967   Capitalism and Underdevelopment in Latin America. New York:
            Monthly Review Press.
    1969   Latin America: Underdevelopment or Revolution. New York: Monthly
            Review Press.
Franke, Richard W.
    1974   Miracle Seeds and Shattered Dreams in Java. Natural History 83 (1):11,
            12, 16, 18, 84–88.
Frankel, Francine R.
    1969   Agricultural Modernization and Social Change. Mainstream, November
            29.
    1971   India's Green Revolution. Princeton, NJ: Princeton University Press.
Fried, Morton
    1960   On the Evolution of Social Stratification and the State. *In* Culture in
            History. S. Diamond, ed. Pp.713–731. New York: Columbia University
            Press.
Geertz, Clifford
    1963   Agricultural Involution. Berkeley and Los Angeles: University of Cali-
            fornia Press.
Gladwin, Christina H.
    1979   Production Functions and Decision Models: Complementary Models.
            American Ethnologist 6(4):653–674.
    1980   A Theory of Real-Life Choices: Applications to Agricultural Decisions.
            In Agricultural Decision Making. Peggy F. Barlett, ed. Pp.45–86. New
            York: Academic Press.
Gladwin, Hugh
    1975   Looking for an Aggregate Additive Model in Data from a Hierarchical
            Decision Process. *In* Formal Methods in Economic Anthropology,
            Stuart Plattner, ed. Pp.159–196. Washington, DC: American Anthropo-
            logical Association.
Gladwin, Hugh, and Michael Murtaugh
    1980   The Attentive/Preattentive Distinction in Agricultural Decision Mak-
            ing. *In* Agricultural Decision Making. Peggy F. Barlett, ed. Pp.115–136.
            New York: Academic Press.
Gleave, M. B., and H. P. White
    1969   Population Density and Agricultural Systems in West Africa. *In* En-
            vironment and Land Use in Africa, M. F. Thomas and G. W. Whitting-
            ton, eds. Pp.273–300. London: Methuen.
Goldkind, Victor
    1961   Sociocultural Contrasts in Rural and Urban Settlement Types in Costa
            Rica. Rural Sociology 26:365–380.
Gonzales-Vega, Claudio
    1973   Small Farmer Credit in Costa Rica: The Juntas Rurales. Country Study
            for AID Spring Review. Vol. 2:1–130. Washington DC: Agency for
            International Development.

Gotsch, Carl H.
  1973   Some Observations on the Small Farmer Credit Problem. *In* Small
         Farmer Credit Summary Papers, A.I.D. Spring Review of Small Farmer
         Credit 20(SR120):79–84. Washington, DC: Agency for International
         Development.
Greenwood, Davydd J.
  1976   Unrewarding Wealth: The Commercialization and Collapse of Agricul-
         ture in a Spanish Basque Town. Cambridge: Cambridge University
         Press.
Griffin, Keith
  1969   Underdevelopment in Spanish America. London: George Allen and
         Unwin.
  1974   The Political Economy of Agrarian Change: An Essay on the Green
         Revolution. Cambridge, MA: Harvard University Press.
Hall, Carolyn
  1978   El Café y El Desarollo Histórico-Geográfico de Costa Rica. San José:
         Editorial Costa Rica y Universidad Nacional.
Halperin, Rhoda
  1977   Introduction: The Substantive Economy in Peasant Societies. *In* Peasant
         Livelihood, Rhoda Halperin and James Dow, eds. Pp.1–16. New York:
         St. Martin's Press.
Halperin, Rhoda, and James Dow, eds.
  1977   Peasant Livelihood: Studies in Economic Anthropology and Cultural
         Ecology. New York: St. Martin's Press.
Hanks, Lucien
  1972   Rice and Man. Chicago: Aldine.
Harner, Michael J.
  1970   Population Pressure and the Social Evolution of Agriculturalists. South-
         western Journal of Anthropology 26(1):67–86.
  1975   Scarcity, the Factors of Production and Social Evolution. *In* Population,
         Ecology, and Social Evolution. Steven Polgar, ed. Pp.123–138. Chicago:
         Aldine.
Harris, Marvin
  1968   The Rise of Anthropological Theory. New York: Crowell.
  1971   Man, Culture, and Nature. New York: Crowell.
  1972   The Human Strategy: How Green the Revolution? Natural History
         81:28–30.
Haswell, Margaret
  1973   Tropical Farming Economics. London: Longman.
Hildebrand, Peter E.
  1977   Socioeconomic Considerations in Multiple Cropping Systems. Guate-
         mala: Instituto de Ciencia y Tecnología Agrícolas. Sector Público Agrí-
         cola.
Jacoby, Erich H.
  1972   Effects of the "Green Revolution" in South and Southeast Asia. Modern
         Asian Studies 6(1):63–69.
James, Preston
  1959   Latin America. New York: Odyssey Press.
Johnson, Allen W.
  1971a Sharecroppers of the Sertao. Stanford, CA: Stanford University Press.
  1971b Security and Risk-Taking among Poor Peasants. *In* Studies in Economic

Anthropology. George Dalton, ed. Pp.144–151. American Anthropological Association Studies No. 7.
1978   Quantification in Cultural Anthropology. Stanford, CA: Stanford University Press.
1980   The Limits of Formalism in Agricultural Decision Research. *In* Agricultural Decision Making. Peggy F. Barlett, ed. Pp.19–44. New York: Academic Press.
Junta de Defensa del Tabaco, Costa Rica
1973   Estudio de Costos de Producción en Tabaco de Sol. Mimeographed.
Knight, C. Gregory
1974   Ecology and Change. New York: Academic Press.
Ladejinsky, Wolf
1973   How Green is the Indian Green Revolution? Economic and Political Weekly 8(52):133–144.
Lange, Frederick
1976   The Northern Central American Buffer: A Current Perspective. Latin American Research Review 11:177–183.
Lee, Richard Borshay
1979   The !Kung San: Men, Women, and Work in a Foraging Society. New York: Cambridge University Press.
Leung, Woot Tsuen
1961   Food Composition Tables for Use in Latin America. Interdepartmental Committee on Nutrition for National Defense. Bethesda, MD: National Institutes of Health.
Lewis, Oscar
1955   Medicine and Politics in a Mexican Village. *In* Health, Culture, and Community. Benjamin D. Paul, ed. Pp.403–434. New York: Russell Sage.
Lipton, Michael
1968   The Theory of the Optimizing Peasant. Journal of Development Studies 4(3):327–351.
1977   Why Poor People Stay Poor. Cambridge, MA: Harvard University Press.
Loomis, Robert S.
1976   Agricultural Systems. Scientific American 235(3):99–104.
Lothrop, Samuel
1926   Pottery of Costa Rica and Nicaragua. New York: Museum of the American Indian, Heye Foundation.
Margolies, Maxine
1977   Historical Perspectives on Frontier Agriculture as an Adaptive Strategy. American Ethnologist 4(1):42–64.
McCay, Bonnie J.
1978   Systems Ecology, People Ecology, and the Anthropology of Fishing Communities. Human Ecology 6(4):397–422.
Meehan, Peter Martin
1978   Staple Theory as a Techno-Economic Theory of Culture Change, with Special Reference to the Rise of Commercial Cattle Production in a Costa Rican Community. M.A. thesis (sociology-anthropology), Iowa State University.
Mencher, Joan P.
1970   Change Agents and Villagers. Economic and Political Weekly 5:29–31. Special Number. July.

1978   Agriculture and Social Structure in Tamilnadu: Past Origins, Present Transformations, and Future Prospects. Durham, NC: Carolina Academic Press.

Mintz, Sidney
1967   Pratik: Haitian Personal Economic Relationships. *In* Peasant Society. Jack M. Potter, May N. Diaz, and George M. Foster, eds. Pp.98–109. Boston: Little, Brown.

Moerman, Michael
1968   Agricultural Change and Peasant Choice in a Thai Village. Berkeley: University of California Press.

Munthe-Kass, Harald
1970a   The Landed and the Hungry. Far Eastern Economic Review 68:27–30.
1970b   Green and Red Revolutions. Far Eastern Economic Review 68:27–30.

Netting, Robert McC.
1968   Hill Farmers of Nigeria. Seattle: University of Washington Press.
1969   Ecosystems in Process: A Comparative Study of Change in Two West African Societies. *In* Contributions to Anthropology: Ecological Essays. David Damas, ed. Pp.102–112. Ottawa: National Museum of Canada Bulletin 230.
1974   Agrarian Ecology. Annual Review of Anthropology 3:21–56.
1977   Cultural Ecology. Menlo Park, CA: Cummings.

Nunley, Robert E.
1960   The Distribution of Population in Costa Rica. Washington, DC: National Academy of Sciences, National Research Council.

Orlove, Benjamin S.
1977a   Integration Through Production: The Use of Zonation in Espinar. American Ethnologist 4(1):84–101.
1977b   Alpacas, Sheep and Men: The Wool Export Economy and Regional Society in Southern Peru. New York: Academic Press.

Ortiz, Sutti
1967   The Structure of Decision-Making Among the Paez Indians of Colombia. *In* Themes in Economic Anthropology. Raymond Firth, ed. Pp.191–228. London: Association of Social Anthropology, Monograph Number 6.
1973   Uncertainties in Peasant Farming. London: Athlone Press.
1976   The Effect of Risk Aversion Strategies on Subsistence and Cash Crop Decisions. Presented at Agricultural Development Council Conference on Uncertainty and Agricultural Development. Mexico City.
1980   Forecasts, Decisions and the Farmer's Response to Uncertain Environment. *In* Agricultural Decision Making. Peggy F. Barlett, ed. Pp.177–207. New York: Academic Press.

Paddock, William C.
1970   How Green is the Green Revolution? Bioscience 20(16):897–902.

Pelto, Pertti J.
1973   The Snowmobile Revolution: Technology and Social Change in the Arctic. Menlo Park, CA: Cummings.

Pelto, Pertti, and Gretel H. Pelto
1975   Intra-Cultural Diversity: Some Theoretical Issues. American Ethnologist 2(1):1–18.

Pelzer, K. J.
1945   Pioneer Settlement in the Asiatic Tropics. New York: American Geographical Society.

Pospisil, Leopold
    1963   Kapauku Papuan Economy. New Haven, CT: Yale University Publications in Anthropology No. 67.
Prothero, R. M.
    1972   People and Land in Africa South of the Sahara. New York: Oxford University Press.
Rappaport, Roy A.
    1968   Pigs for the Ancestors. New Haven, CT: Yale University Press.
Rawski, Evelyn Sakakida
    1972   Agricultural Change and the Peasant Economy of South China. Cambridge, MA: Harvard University Press.
Redfield, Robert
    1950   A Village that Chose Progress. Chicago: University of Chicago Press.
Richerson, P. J.
    1977   Ecology and Human Ecology: A Comparison of Theories in the Biological and Social Sciences. American Ethnologist 4(1):1–26.
Rochin, Refugio I.
    1977   Rural Poverty and the Problem of Increasing Food Production on Small Farms: The Case of Colombia. Western Journal of Agricultural Economics 1(1):181–186.
Romanucci-Ross, Lola
    1973   Conflict, Violence, and Morality in a Mexican Village. Palo Alto, CA: National Press Books.
Roumasset, James A.
    1979   Unimportance of Risk for Technology Design and Agricultural Development Policy. In Economics and the Design of Small-Farmer Technology. Alberto Valdes et al., eds. Pp.48–65. Ames, IA: University of Iowa Press.
Roumasset, James A., Jean-Marc Boussard, and Inderjit Singh, eds.
    1979   Risk, Uncertainty, and Agricultural Development. New York: Agricultural Development Council.
Ruthenberg, Hans, ed.
    1968   Smallholder Farming and Smallholder Development in Tanzania. Munich: Weltforum Verlag.
Ruttan, Vernon W., and Hans P. Binswanger
    1978   Induced Innovation and the Green Revolution. In Induced Innovation: Technology, Institutions, and Development. Hans P. Binswanger and Vernon W. Ruttan, eds. Pp.358–408. Baltimore: Johns Hopkins University Press.
Sahlins, Marshall D.
    1964   Culture and Environment: The Study of Cultural Ecology. In Horizons in Anthropology. Sol Tax, ed. Pp.132–147. Chicago: Aldine.
Sanders, William T., and Barbara J. Price
    1968   Mesoamerica: The Evolution of a Civilization. New York: Random House.
Sandner, Gerhard
    1960   Turrubares: Estudio de Geografía Regional, Problemas Sociales de la Expansión Agrícola en Costa Rica. San José: Instituto Geográfico de Costa Rica.
Schultz, Theodore W.
    1945   Food for the World. Chicago: University of Chicago Press.

1964 Transforming Traditional Agriculture. New Haven, CT: Yale University Press.
Seligson, Mitchell A.
1975 Agrarian Capitalism and the Transformation of Peasant Society: Coffee in Costa Rica. Buffalo, NY: Council on International Studies, SUNY Buffalo Special Studies 69.
1980 Peasants of Costa Rica and the Development of Agrarian Capitalism. Madison, WI: University of Wisconsin Press.
Service, Elman
1955 Indian-European Relations in Colonial Latin America. American Anthropologist 54:411–455.
Skorov, Georgy
1973 The Green Revolution and Social Progress. World Development 1(11):13–21.
Smith, Carol A.
1978 Beyond Dependency Theory: National and Regional Patterns of Underdevelopment in Guatemala. American Ethnologist 5(3):574–617.
Smith, Waldemar R.
1977 The Fiesta System and Economic Change. New York: Columbia University Press.
Spooner, Brian
1972 Population Growth. Cambridge, MA: MIT Press.
Steward, Julian
1948 Handbook of the South American Indians, Vol 5. Washington, DC: Smithsonian Institution. U.S. Government Printing Office.
Stone, Samuel Z.
1975 La Dinastia de los Conquistadores: La Crisis del Poder en la Costa Rica Contemporánea. San José: Editorial Universitaria Centroamericana.
Strickon, Arnold, and Sidney M. Greenfield, eds.
1972 Structure and Process in Latin America: Patronage, Clientage, and Power Systems. Albuquerque: University of New Mexico Press.
Takahashi, Akira
1970 Land and Peasants in Central Luzon. Honolulu: East-West Center Press.
Thein, Tin Myaing
1975 Contraception in Costa Rica: A Sociocultural Study of Three Communities. Ph.D. dissertation (anthropology), Columbia University.
U.S. Department of Commerce, Bureau of the Census
1973 Statistical Abstract of the United States. Washington, DC: Bureau of the Census. U.S. Government Printing Office.
Van Vekken, P. M., and H. V. E. Thoden Van Velzen
1972 Land Scarcity and Rural Inequality in Tanzania. The Hague: Mouton.
Vayda, Andrew P., ed.
1969 Environment and Cultural Behavior. Garden City, NY: Natural History Press.
Vayda, Andrew P., and Bonnie J. McCay
1975 New Directions in Ecology and Ecological Anthropology. Annual Review of Anthropology 4:293–306.
West, Robert C.
1957 The Pacific Lowlands of Colombia. Baton Rouge, LA: Louisiana State University Studies Social Science Series 8.

Wharton, Clifton R., Jr.
  1969   The Green Revolution: Cornucopia or Pandora's Box? Foreign Affairs
         (April), pp.464–476.
  1971   Risk, Uncertainty, and the Subsistence Farmer: Technological Innova-
         tion and Resistance to Change in the Context of Survival. *In* Studies in
         Economic Anthropology. George Dalton, ed. Washington, DC: Amer-
         ican Anthropological Association, Anthropological Studies #7:152–179.
White, Benjamin
  1973   Demand for Labor and Population Growth in Colonial Java. Human
         Ecology 1(3):217–244.
White, Leslie
  1959   The Evolution of Culture. New York: McGraw-Hill.
Whitten, Norman E., Jr., and Dorothea S. Whitten
  1972   Social Strategies and Social Relationships. Annual Review of Anthropol-
         ogy 1:247–270.
Wilber, Charles K., and James H. Weaver
  1979   Patterns of Dependency: Income Distribution and the History of Under-
         development. *In* The Political Economy of Development and Underde-
         velopment. Charles K. Wilber, ed. Pps.114–129. New York: Random
         House.
Wilkinson, Richard G.
  1973   Poverty and Progress: An Ecological Perspective on Economic Develop-
         ment. New York: Praeger.
Winkelmann, Donald
  1976   The Adoption of New Maize Technology in Plan Puebla, Mexico.
         Mexico City: CIMMYT.
Wolf, Eric
  1955   Types of Latin American Peasantry. American Anthropologist
         57(3):452–471.

Index

DATE DUE

2 07 '85	
7 25 '85	
1 21 '87	
15 28 '88	
10 21 '87	
11 25 '87	
NOV 2 2 '88	
Pet 1 - 3 - 90	
APR 2 1 1995	
APR 1 1996	
NOV 1 6 1996	